Dielectrics 2009: Measurement Analysis and Applications, 40th Anniversary Meeting

Journal of Physics: Conference Series Volume 183

Reading, Berkshire, United Kingdom
15-17 April 2009

ISBN: 978-1-61738-301-4

Printed from e-media with permission by:

Curran Associates, Inc.
57 Morehouse Lane
Red Hook, NY 12571
www.proceedings.com

Some format issues inherent in the e-media version may also appear in this print version.

Printed by Curran Associates, Inc. (2010)

Institute of Physics
Dirac House, Temple Back
Bristol BS1 6BE UK

Phone: 44 1 17 929 7481
Fax: 44 1 17 920 0979

techtracking@iop.org

TABLE OF CONTENTS

PREFACE

The conference Dielectrics 2009: Measurements, Analysis and Applications represents a significant milestone in the evolution of dielectrics research in the UK. It is reasonable to state that the academic study of dielectrics has led to many fundamental advances and that dielectric materials underpin the modern world in devices ranging from field effect transistors, which operate at extremely high fields, albeit low voltages, to the high voltage plants that provide the energy that powers our economy. The origins of the Dielectrics Group of the Institute of Physics (IOP), which organized this conference, can be traced directly back to the early 1960s, when Professor Mansel Davies was conducting research into the dielectric relaxation behaviour of polar liquids and solids at The Edward Davies Chemical Laboratories of the University College of Wales, Aberystwyth. He was already well-known internationally for his studies of molecular structure and bonding of small molecules, using infra-red-spectroscopy, and of the physical properties of hydrogen-bonded liquids and solids, using thermodynamic methods. Dielectric spectroscopy was a fairly new area for him and he realized that opportunities for scientists in the UK to gather together and discuss their research in this developing area of physical chemistry/chemical physics were very limited. He conceived the idea of forming a Dielectrics Discussion Group (DDG), which would act as a meeting point and provide a platform for dielectrics research in the UK and beyond and, as a result, a two-day Meeting was convened in the spring of 1968 at Gregynog Hall of the University of Wales, near Newtown, Montgomeryshire. It was organized by Mansel Davies, Alun Price and Graham Williams, all physical chemists from the UCW, Aberystwyth. Fifty scientists attended, being a mix of physical chemists, theoretical chemists, physicists, electrical engineers, polymer and materials scientists, all from the UK, except Dr Brendan Scaife of Trinity College, Dublin. The Meeting discussed dielectric relaxation behaviour arising from molecular motions of dipolar molecules in the liquid and elastomeric states (now known as soft condensed matter) with measurements spanning a frequency range from a few Hz, through power and radiofrequencies, UHF and VHF, the microwave range and into the far infra-red. As a result of its success, it was decided at the Meeting that a continuing Dielectrics Discussion Group would be established to meet not more frequently than once a year.

It was appreciated at the time that the subject of 'Dielectrics' covered many sub-areas, broadly classified into those of polarization, relaxation, conduction and high-field phenomena. For the DDG, a solution was to run annual meetings on chosen themes in dielectrics research, where the theme would change from one meeting to the next. Topics addressed in the early years of the DDG included high field phenomena and impurity effects, heterogeneous systems and biomaterials, polarization and conduction and non-linear dielectrics and ferroelectrics. The number of participants at these early meetings grew from 50 to the low 100's, which reflected both the increased awareness of the Group and increased participation from researchers in Continental Europe. However, the majority of participants in this period were from the UK, which reflected the considerable activities in dielectrics research in University, Industry and Government laboratories in the UK. There followed a series of DDG Meetings until 1974, at which point, the DDG became a registered charity: The Dielectrics Society. Many of the earlier meetings were held in the attractive and convenient venues of Oxford and Cambridge colleges but, in the early 1990's, a new venue for the meetings was established at the University of Kent at Canterbury. In 2001, the next major change occurred when The Dielectrics Society was incorporated into the Institute of Physics, becoming their Dielectrics Group.

From 1968 to 2001 the Annual Meetings focused on numerous topics, including relaxation and conduction processes in liquids, solids, liquid crystals, synthetic polymers and biopolymers, piezoelectric materials, electrets and ferroelectrets, interfacial phenomena, high field conduction and breakdown phenomena in solids, liquids and gases and, importantly, the remarkable developments in dielectric instrumentation during this period. These activities reflected the need, and willingness, to move dielectrics researches with the times. As examples of the variety and diversity of these meetings we may refer briefly to the 1981, 1989 and 1996 Meetings. The 1981 Oxford Meeting on High Field Phenomena in Dielectrics included strong themes on fundamental and practical effects of high E-fields on the dielectric and conduction behaviour of liquids and solids, electrical treeing and dielectric breakdown, non-linear dielectric effects, electrets, thin-film devices and electro-rheology. The late 1980's had seen large initiatives in the UK and globally in the general area of Molecular Electronics so, in timely fashion, this was the subject of the 1989 Meeting in Bangor. The 1996 Smart Dielectrics Meeting at Canterbury reported subsequent advances in designer materials having electro-responsive and electro-optical properties. The programme concerned electro- and photo-active materials, mainly organic, in the form of polar dielectrics, polyelectrolytes, organic semi- and photo-conductors, photo-refractive polymer films, organic ferroelectric films, liquid crystalline polymer films, piezo- and pyro-electric polymer films, electroluminescent polymers, electro-rheological fluids and non-linear optical polymer films as described by leading international scientists. The physico-chemical functions of the materials were demonstrated and interpreted in terms of fundamental molecular properties. An Archive, containing full details of all the Meetings of the DDG and the Dielectrics Society, has been placed on the Website of the IOP Dielectrics Group by Professor John Fothergill.

The 2009 meeting of The Dielectrics Group of the IOP at Reading was therefore held to celebrate 40 years of coordinated dielectrics research activity in the UK and beyond, as presented at these DDG and Dielectrics Society Meetings. To reflect the range of topics addressed since the first DDG meeting in 1968, the organizing committee chose to break away from the tradition of a stronglythemed event and, rather, adopted a general, inclusive approach, albeit based upon four broad technical areas:

Theme 1: Molecular and bulk relaxation processes
Theme 2: Space charge and charge transport in insulators
Theme 3: Functional materials
Theme 4: Bio-dielectrics and complex systems.

The result was a highly successful conference that attracted some 90 delegates from 11 countries, giving the event a truly international flavour, and which included both regular and new attendees. Consequently, the organizing committee would like to thank, our colleagues at the IOP, the invited speakers, our sponsors at National Grid, EDF Energy and Dow Wire and Cables and all the delegates for making the event such a success.

Professor Alun Vaughan, Professor Graham Williams

Journal of Physics: Conference Series **183** (2009) 012001 doi:10.1088/1742-6596/183/1/012001

Use of scanning LIMM (Laser Intensity Modulation Method) to characterise polarisation variability in dielectric materials

Mark Stewart and Markys Cain

National Physical Laboratory, Hampton Road, Teddington, Middlesex, TW11 0LW, UK

e-mail: mark.stewart@npl.co.uk

Abstract. The Laser Intensity Modulation Method (LIMM) has traditionally been used to characterise the depth dependence of polarisation of piezoelectric materials[1]. Although the technique is simple, it is difficult to extract the polarisation / depth data from the measured pyroelectric current because of the complex mathematics pertaining to the physics of the technique. However, the laser probe may still be used as a comparative or qualitative tool in mapping out the polarisation across the surface of a material. A novel scanning LIMM system has been developed to map the variation in piezoelectric activity across a range of samples. The system has been upgraded with a galvanometer mirror scanner to increase speed and reduce sensitivity to acoustic noise. The improvements are discussed and tested on a range of case studies. The technique can be used to show differences in piezoelectric activity caused by features such as inhomogeneous material composition, porosity and mechanical damage. The method has application as a quality control tool for materials and device manufacturers.

1. Introduction

The laser intensity modulation method (LIMM) was developed in the early 1980's as a method to characterize the through thickness polarization of polymeric ferroelectric thin films[1, 2]. In this method an intensity modulated radiation source is used to generate a harmonic thermal wave in a sample, via an absorbing electrode, which in turn generates a pyroelectric current in the material. Because of the time constant of this thermal wave it can sample to different depths, depending on the frequency of modulation and the thermal diffusivity of the material. Most work has concentrated on quantifying the results of the experiment[3-5], and because of the complexity of this task workers have tended to use the method in a one-dimensional mode. Recently the method has been used in a more qualitative manner to get three-dimensional information from the technique[6-9]. In the scanning LIMM method a focused laser beam must be moved across the test sample surface, and until now this has been achieved by moving the sample on an x-y table. This traveling stage can introduce acoustic noise into the measurements, particularly when examining piezoelectric ceramic samples with a relatively high mechanical quality factor. In this work we report for the first time the use of galvanometer mirror scanners to scan the beam across the sample.

The generation of current in the scanning LIMM technique is also more involved than in the one-dimensional case. In general, thermally generated currents are termed pyroelectric currents, and for a ferroelectric material the measured pyroelectric coefficient is given by the following[10]-

$$p_i^{\sigma} = p_i^{\varepsilon} + \alpha_{jk}^{E} c_{jklm}^{E,T} d_{ilm}^{T} \tag{1}$$

where the first term is the pyroelectric current measured at constant strain, termed the *primary* pyroelectric effect, p_i^{ε}, and the second term is current produced from piezoelectric effects via thermal expansion. In the scanning LIMM technique there are further possibilities, such as where thermal expansion in an inactive part of the sample causes a thermoacoustic wave, which is then picked up piezoelectrically in the sample. All these mechanisms can lead to a signal, so quantitative evaluation is difficult, nevertheless as will be shown the technique can produce useful qualitative data.

The harmonic heat flux problem can be solved analytically for the one-dimensional case, where the laser beam diameter is much broader than the penetration depth, to give the temperature at any depth[3]. From this it can be shown that the penetration depth of the thermal wave, x, depends on the angular frequency, ω, and the thermal diffusivity, α, as follows

$$x = \sqrt{2\alpha/\omega} \tag{2}$$

The thermal diffusivity of the material studied here, Lead Zirconate Titanate, PZT, is 4.5 x 10^{-7} m^2/s [11], which means the wave will penetrate 0.38mm at a frequency of 1Hz and only 3.78 μm at a frequency of 10kHz

2. Experimental

A schematic of the LIMM system developed at NPL is shown in figure 1. The system is based around a 785nm, 75mW diode laser with an analogue input to control laser intensity that is capable of modulation frequencies of up to 1MHz. The analogue modulation is controlled by the oscillator output of a Stanford Research Systems SR830 Lock-in amplifier, which is also used to acquire the pyroelectric current. A FEMTO DLPCA-200 current to voltage converter is used to process the pyroelectric signal before it is sent to the lock-in amplifier. The sample stage is a stepper motor, Danaher Precision x-y stage, capable of 50mm travel with a step size of 1 μm. In the present system the laser is mounted horizontally on the side of the case and the beam is deflected through 90° by the x-y mirror scanner before reaching the sample. The scanners are Cambridge Technology 6210 galvanometer scanners with 3mm mirrors, and controlled by a PositionPro-2 hardware controller. The

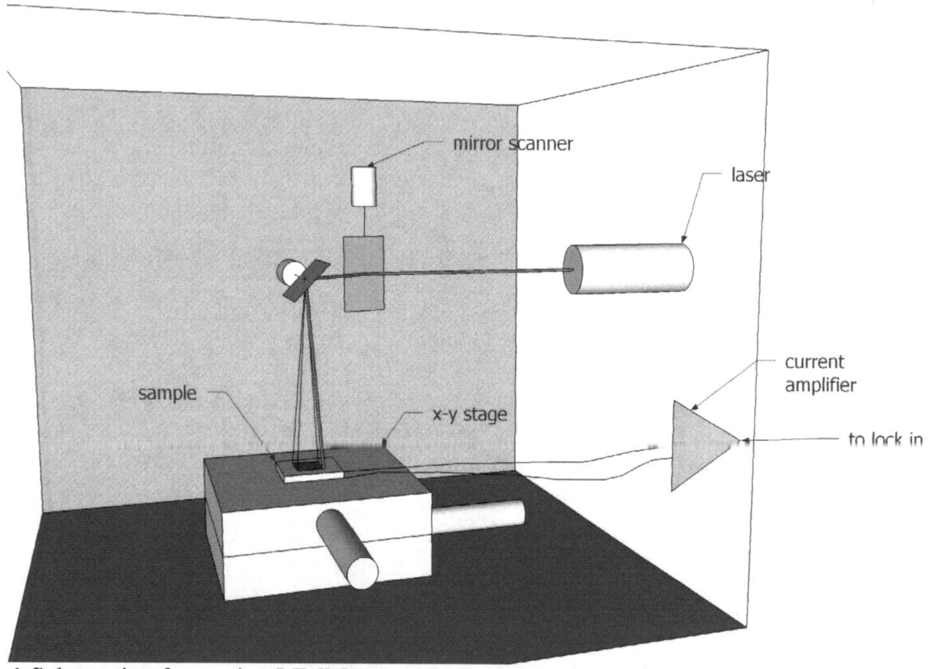

Figure 1 Schematic of scanning LIMM system.

angular range of the scanners is ±20degrees, with a step size of 1.14 μm. The instrument enclosure is a metal box that is interlocked for the purpose of laser safety, and also acts as a Faraday cage to mitigate the influence of external electromagnetic fields on the sensitive current measurements. The system is mounted on vibration dampening rubber feet, but as the size of the box has been increased to accommodate the necessary components it has become more susceptible to airborne noise. This aspect of the performance could be improved by adding an extra acoustic isolation layer around the enclosure.

3. Results and Discussion

3.1. Thermal Modelling

To help to understand the physical processes in the LIMM experiment, a 1-D thermal finite volume model, TherMOL[12], was constructed for a 0.15mm thick sample of a PZT material illuminated with a laser. Figure 2 shows the modeled temperature rise at two depths on the sample, the front surface where the laser is incident, and the back surface. These are also shown for two different angular frequencies of the intensity modulation of the incident laser radiation. The incident laser is modeled as a sinusoidal heat flux, which is applied as a boundary condition, that is, all the laser radiation, (modified by a suitable absorption coefficient) is converted to heat in an infinitely thin layer on the top surface. In reality this may not be the case since this depends on the transparency of the absorbing electrode to the incident radiation. The temperature rise on the front surface shows two effects, a slow temperature rise which will eventually reach a steady state, and on top of this a higher frequency temperature modulation at the frequency of the incident radiation. The slow temperature rise is independent of the modulation frequency as it is based on the average laser intensity, and although this does give rise to a pyroelectric charge this component is removed by use of lock in techniques. The higher frequency component is measured using the lock-in and for a particular frequency and other experimental conditions this is constant over time. This implies that in the experiment the material under test should not have temperature dependent, or time dependent properties. The temperature rise at the back surface also shows a slow temperature rise, but for the low frequency case the amplitude is

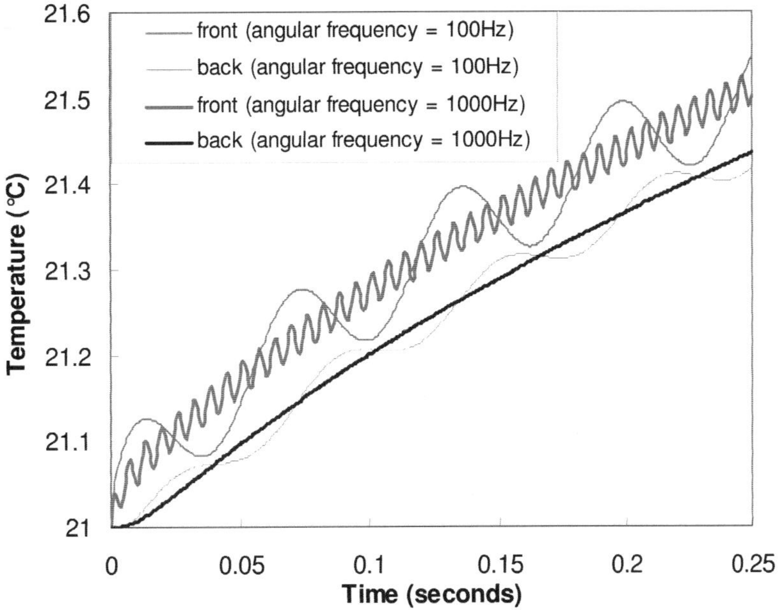

Figure 2 Finite volume 1D thermal model of temperature rise of 150 μm thick PZT sample with a harmonic heat flux applied to the front surface.

reduced and there is a large phase shift, whilst in the high frequency case there is no visible modulation. This illustrates the depth sensitivity of the LIMM technique, at high frequencies the signal is generated from the top surface, and reducing frequency samples through more of the thickness.

3.2. Scanning System

One of the original aims of adding the mirror scanners to the system was to reduce the sensitivity to acoustic noise pickup. Figure 3 shows the pyroelectric current over time whilst moving either the x-y stage or the beam every ten seconds. Moving the x-y stage generates a sharp pulse in the signal every time the stage moves, whilst moving the beam with the mirrors does not. Secondly there is additional noise generated by the x-y stage due to the energizing of the stepper motor coils needed to maintain position. This can be seen when the motors are turned off after forty seconds in the 'move stage' experiment, and is entirely missing from the 'move beam' experiment as the motors were off.

Although it would appear that by changing to moving the beam, rather than moving a mechanical stage, the acoustic noise of the stage has been eliminated, the use of the mirror scanners has revealed a further artifact of the scanning LIMM method. The effect, which is similar to the acoustic noise problem, is a consequence of a pyroelectric signal only being generated by a *change* in temperature. This implies a constant beam focused in one spot, will after thermal equilibrium is reached, not generate any further current. If however the focused beam is moved to another position, as in a scanning system, then the original spot will see a sharp drop in temperature and the new position a sharp rise leading to a current. As they are opposite in sense they will tend to cancel out, but this effect has been seen in scanning LIMM measurements and is most prominent when the step size of the scan is larger than the thermal diffusion zone of the beam.

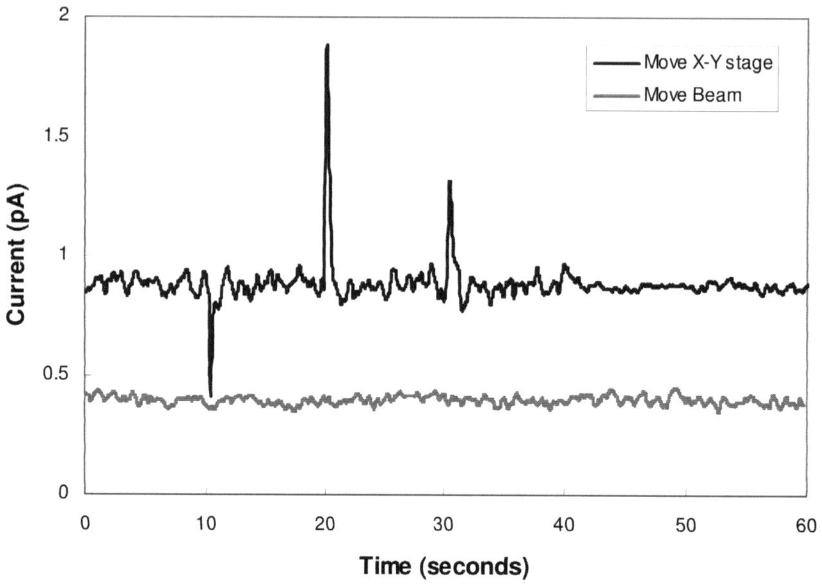

Figure 3 Pyroelectric signal from PZT sample whist moving the motorized stage (blue line) or moving the beam (red line). The motor was powered off after 40 seconds for the stage move.

Overall the addition of the mirror scanner has led to an improvement in the speed of the system of roughly 5 fold. Typically in the previous system when moving the stage it was necessary to wait for 500ms after each step to be sure that any transient noise in the system had time to settle. In theory moving the stage at a constant velocity and matching the acquisition time accordingly could reduce this transient. However as has been shown, even energizing the stage introduces a small amount of noise which is not seen when using the mirror scanners.

3.3. Scanning Results

To illustrate the use of the system a series of scans were performed on typical piezoelectric materials and devices.

3.3.1. Piezo Composite Material.

Figure 4 shows a scanning LIMM image of a piezo composite sample. The composite consists of 250μm diameter piezoelectric fibres embedded in an epoxy matrix. The regions of high pyroelectric activity are piezoelectric fibres seen end-on, whilst the epoxy is inactive. The scanned images consist of 80 by 80 points at a step size of 50μm and took approximately 30 minutes to acquire. The depth sensitivity of the technique can be seen in the different frequency images, one at 1kHz, which samples approximately 12μm and one at 100Hz which samples to a depth of approximately 38μm, with the sample thickness being 1mm. The lower frequency image shows the piezoelectric fibres more clearly, and some are not visible at all in the high frequency image. This could be because either the fibre is not piezoelectrically active, or more likely that there is a thin layer of epoxy covering the face of the fibre that the high frequency thermal wave cannot penetrate. The sample has gold electrodes both top and bottom which are needed first to pole the sample and then later to use the device. One of the issues with piezocomposite materials is the difficulty of generating a field across the piezoelectric material in the presence of an insulating polymer. This technique offers a way of examining which areas of the sample have had a field high enough to pole the piezoelectric.

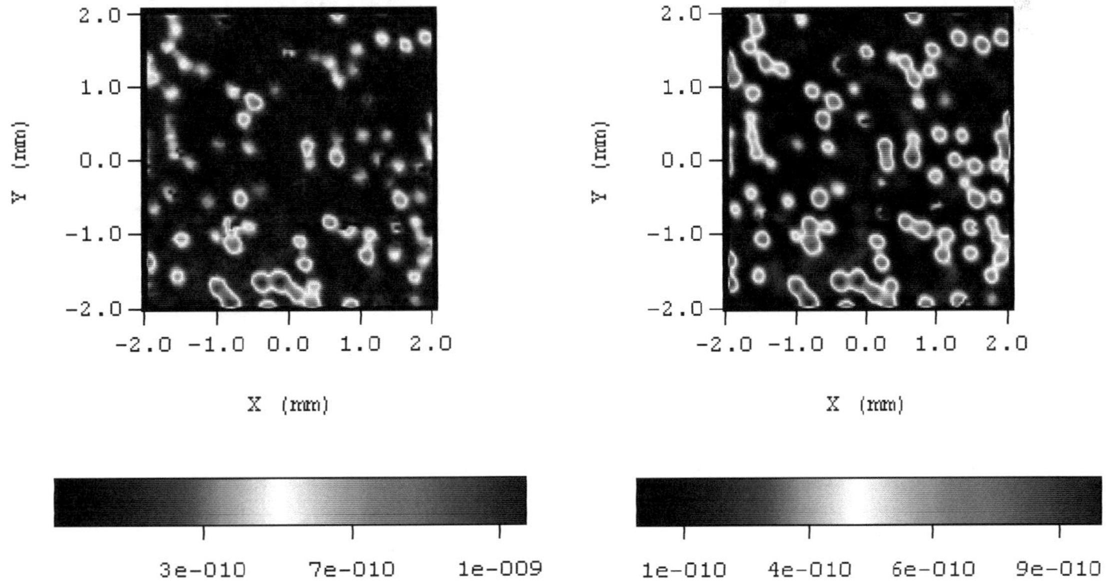

Figure 4 Scanning LIMM image of 1-3 composite material, left hand image at 1kHz, right hand image at 100Hz. Z scale is current in Amps.

3.3.2. Piezoelectric multilayer with artificial defects.

In a piezoelectric multilayer layers of piezoelectric material are alternately sandwiched with metal electrodes in order to reduce the operating voltage whilst still maintaining a high field.Figure 5 shows the top surface of a 15mm square, 3mm thick, piezoelectric multilayer which has model defects introduced in the form of 2.5mm diameter circles patterned as the number five on a dice. The piezoelectric layers are stacked in the plane of the image and the electrical contacts to the layers are taken off at the left and right of the sample. In this case the LIMM image is formed where the top absorbing layer acts only as a light to thermal converter, and does not form the collecting electrode. The material is slightly transparent in the visible spectrum and so not all of the heat is produced directly at the top surface. The two images taken at 100Hz and 1Hz clearly show the reduced output from the defect areas, and it can be seen that the defect layers on alternate layers are offset left to right in the direction of the electrical contacts. This is because each metal layer must stop short of going across the full width of the sample otherwise it will short circuit. In the higher frequency image the circles are slightly more defined whereas in the low frequency image they tend to blur together. The high frequency image shows the right hand circles as slightly larger and appear more "in-focus", as this layer is nearer to the top surface. The piezoelectric induced displacement of these multilayers has already been investigated[13], and it is interesting to note that the patterns seen were elongated ellipses, as in the low frequency LIMM image, and it was not possible to distinguish two separate layers. This is because the displacement generated comes from the entire thickness of the multilayer, rather than just a few layers in the scanning LIMM results.

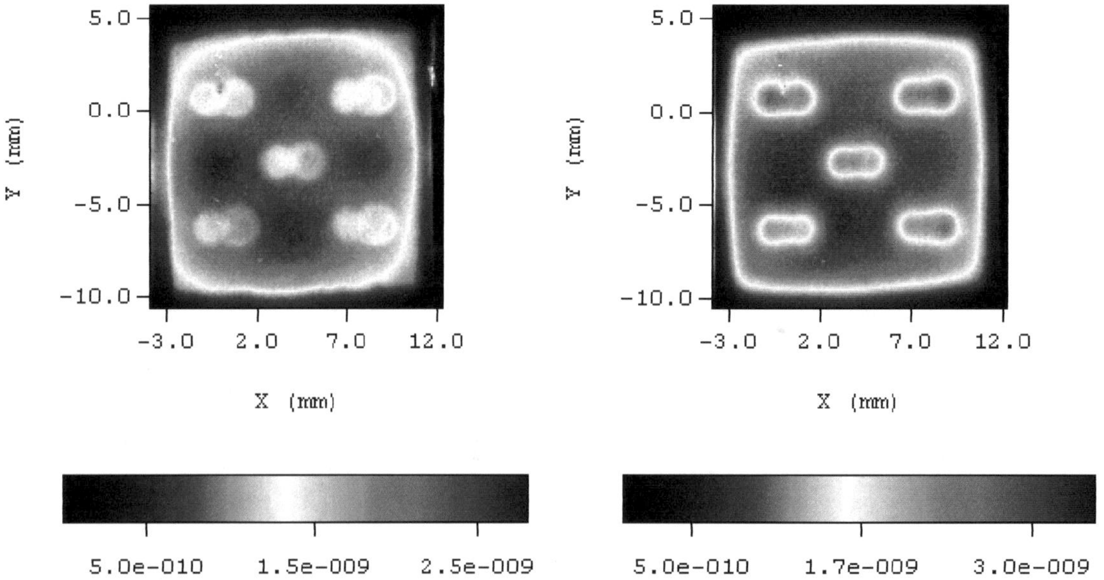

Figure 5 Scanning LIMM image of piezoelectric multilayer samples, left hand image at 100Hz, right hand image at 1Hz. Z scale is current in Amps.

3.3.3. Piezoelectric multilayer stack.

Figure 6 shows a scanning LIMM image of a piezoelectric stack, where several individual 2.2mm thick chips have been stacked together, with the electrodes being taken off top and bottom of the image. The image is taken from the side of the stack, and again there is no absorbing electrode, simply a capping layer of PZT for electrical isolation purposes. The individual stacks can be identified and they would appear to show different levels of activity. At the right hand side there is an extra PZT chip without electrodes and this gives almost no signal, whereas the left hand one gives the highest output. This is probably due to the differences in the proximity of the electrodes to the surface rather than a real difference in activity of the piezoelectric material. The individual layers of the multilayer cannot be identified as by the time the thermal wave has reached this depth it has already passed through the capping layer thus degrading the resolution. The image also shows some spots of higher activity at the top and bottom between each stack, (the width of the piezoelectric part of the stack is 7mm). This is an artifact where there are regions of solder on the sample and there is a good thermal and mechanical coupling to the rest of the sample leading to a phantom signal.

4. Conclusions

The scanning LIMM system developed at NPL has been upgraded to include an x-y galvanometer mirror scanner in order to scan the beam across the sample, rather than move the sample with an x-y stage. This has led to improved scanning speeds, with a typical step speed of the scanner of less than

1msec, but more importantly less acoustic noise. Overall this has given roughly a 5-fold increase in the speed of the system.

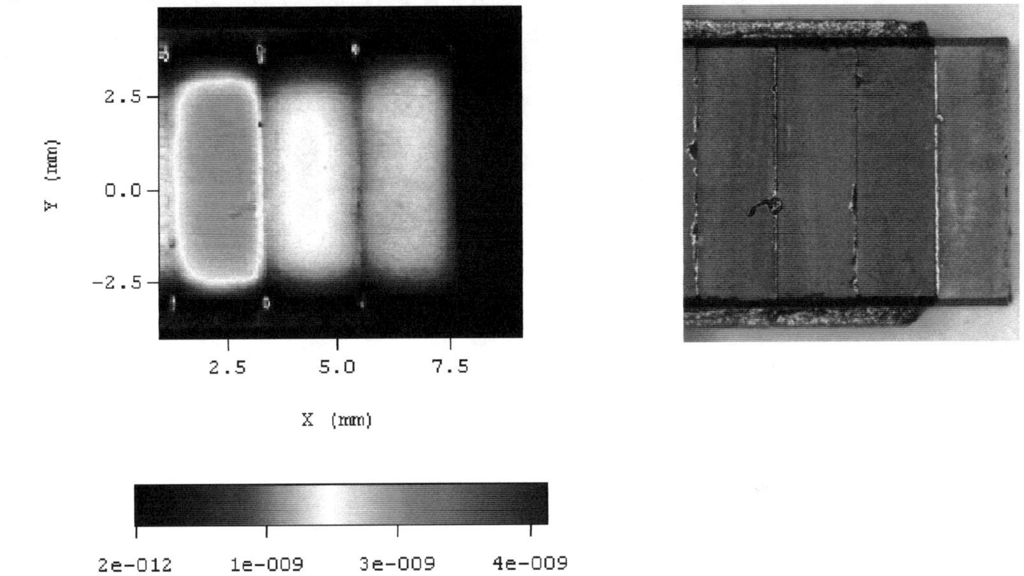

Figure 6 Scanning LIMM image of piezoelectric multilayer sample at 100Hz and corresponding photograph.

The system has been used to examine a broad range of piezoelectric devices and has been demonstrated to show excellent x-y spatial sensitivity and also the depth sensitivity has been used to highlight differences in samples. Conventionally the LIMM method needs an additional radiation absorbing electrode layer, however it has been demonstrated that as long as there are electrodes present in the device for current collection, then this extra step is not needed. This is useful in practical terms if the technique is to be used as a rapid investigative tool.

5. References

[1] Lang S B and D K Das-Gupta, 1981 *Ferroelectrics* **39**(1) 1249 - 1252.
[2] Lang S B and D K Das-Gupta, 1986 *Journal of Applied Physics* **59**(6) 2151-2160.
[3] Ploss B, R Emmerich, and S Bauer, 1992 *Journal of Applied Physics* **72**(11) 5363-5370.
[4] Lang B S, 2004 *Dielectrics and Electrical Insulation, IEEE Transactions on* **11**(1) 3-12.
[5] Lang S B, 1991 *Ferroelectrics* 343-361.
[6] Marty-Dessus D, et al., 2002 *Journal of Physics D: Applied Physics* **35**(24) 3249-3256.
[7] Mellinger A, et al., 2005 *Applied Physics Letters* **86**(8) 082903-3.
[8] Quintel A, J Hulliger, and M Wubbenhorst, 1998 *J. Phys. Chem. B* **102**(22) 4277-4283.
[9] Stewart M and M Cain, 2008 *Journal of the American Ceramic Society* **91**(7) 2176-2181.
[10] Nye J F, *Physical Properties of Crystals: Their representation by tensors and matrices.* 1985, London: Oxford University Press.
[11] Stewart M and M G Cain, in *NPL Measurement Good Practice Guide No. 84.* 2005.
[12] Chapman L A, A T Fry, and S J Roberts, in *NPL Report DEPC-MPE 018.* 2005.
[13] Stewart M, M G Cain, and W Battrick. in *Dielectric Materials, Measurements and Applications, 2000. Eighth International Conference on (IEE Conf. Publ. No. 473).* 2000.

Acknowledgements
This work was supported by the DIUS NMS innovation R&D programme, and the EU F6 Network of Excellence 'MIND' and the Piezo Institute is gratefully acknowledged.

Dielectrics 2009: Measurement Analysis and Applications, 40th Anniversary Meeting IOP Publishing
Journal of Physics: Conference Series **183** (2009) 012002 doi:10.1088/1742-6596/183/1/012002

Dielectric spectroscopy measurements on very low loss cross-linked polyethylene power cables

Tong Liu, John Fothergill, Steve Dodd, Ulf Nilsson*

University of Leicester, UK, LE1 7RH
**Borealis AB, SE-444 86 Stenungsund, Sweden*

tl57@leicester.ac.uk

Abstract. The principles of dielectric spectroscopy are reviewed and the techniques in both time and frequency domains are explored in search of appropriate methods for measurement on low loss XLPE cables. By combining the techniques of frequency response analyzer, transformer ratio bridge and discharging current measurements, some preliminary tests results on homopolymer XLPE model cables have been presented and analyzed, in a wide frequency range of 10^{-4}Hz~2×10^{4}Hz. Dielectric loss mechanisms of XLPE cables are discussed based on the measurement results.

1. Introduction

High voltage cables are widely used to convey electrical power. Research and development of the electrical insulation of power cables are important for improved performance and reliability. Oil-paper insulated power cables were invented by Ferranti in 1891, but modern cables employ polymeric materials as the primary insulation such as polyethylene (PE), cross-linked polyethylene (XLPE) and ethylene propylene rubber (EPR). [1]. Nowadays the insulation system of power cables needs to withstand extremely high voltages of up to 1000kV with reliable long-term operation and with insulation thickness as thin as possible to minimise manufacturing costs. Any insulation defects present in power cables can cause insulation degradation and subsequent electrical breakdown. Because of XLPE's intrinsic breakdown strength of up to 800kV/mm and enhanced melting temperature from 75°C to 90°C [2], XLPE insulation, is the most widely used insulation in high voltage cables. Therefore, in the present study, the dielectric properties of XLPE based cable insulation systems will be used to study the mechanisms of dielectric loss.

Dielectric spectroscopy is a technique used to study the interaction of a material and the applied electric field. It is widely used a tool for the detection of material ageing and fault diagnosis for insulation systems including power cables and hence it has become a popular and powerful research technique [3]. Although various techniques can be used for dielectric spectroscopy, including measurements in the frequency and time domains, the dielectric loss measurement of XLPE cables is usually beyond the abilities of many commercial instruments due to the very low dielectric loss exhibited by this class of material. The low loss of XLPE arises due to its non-polar molecular structure, low levels of impurity and additives. Direct measurements on cable systems impose more difficulty owing to the much larger insulation thickness compared with thin film samples that are often used in laboratory tests, making it difficult to set up the high electrical fields necessary for sensitive dielectric measurements for a given applied voltage. In addition, measurements on thin film samples may not be representative due to differences in the local morphology of the semicrystalline materials obtained under cable manufacturing conditions and those present for the formation of thin films. Also in typical cables, semiconducting shields, which may also contribute to the dielectric loss of the cable, are extruded along with the cable insulation and it would be difficult to replicate the concentration and distribution of the shields and of the crosslinking by-products and impurities using thin film test specimens. For this reason, triple extruded model cables having reduced insulation thickness are used in this study. The dielectric loss mechanisms will be studied by means of the complementary use of frequency and time domain dielectric spectroscopy, in order to cover the widest possible frequency range with the required sensitivity.

1.1. Principle of dielectric spectroscopy

Dielectric spectroscopy is based on the phenomena of electrical polarization and electrical conduction in materials. There are a number of different dielectric polarization mechanisms operating at the molecular or microscopic level. Each polarization mechanism, either a relaxation or resonance processes, is centred around its particular characteristic frequency, which is the reciprocal of the characteristic time of the process and therefore separable in frequency.

The most common mechanisms can be divided into three main categories, as shown in Figure 1. At the highest frequencies, the electric field will cause a slight displacement of the electrons of any atom with respect to the positive nucleus, electronic polarization, while at lower frequencies, atomic polarization is due to the distortion of the arrangement of atomic nuclei in a molecule or lattice. All polymeric materials have these two types of high frequencies polarization and which occur above infra-red frequency. Orientational polarization occurs when particular molecular groups exhibiting a permanent dipole moment, initially orientated randomly in space, tend to be aligned by the applied field to give a net polarization in that direction. The rate of dipolar orientation is highly dependent on inter- and intra-molecular interaction. Orientation of molecular dipoles can therefore occur over a wide range of frequency dependent on the ease with which the dipoles can rotate. Dipoles due to absorbed moisture will be much more easily rotated than polar groups associated with the main polymer chains which may require the co-operative motion of surrounding molecular chains for orientation. For this reason, dielectric spectroscopy ideally suited for the identification and differentiation of various polar groups on main chain or side chains. At very low frequencies, DC conduction will become significant and usually manifests as a slope of -1 in the imaginary permittivity, whilst the real part remains constant. Thus dielectric spectroscopy is also well suited for the determination of DC conductivity of materials. Other common dielectric responses include the quasi-DC mechanism where the low frequency permittivity, real and imaginary, have the same slope resulting from partially mobile charge carriers gradually moving or hopping within the material. Interfacial polarisation (Maxwell-Wagner polarization) can also manifest at low frequency and in this case the real permittivity has a slope of -2 while the imaginary part has slope of -1.

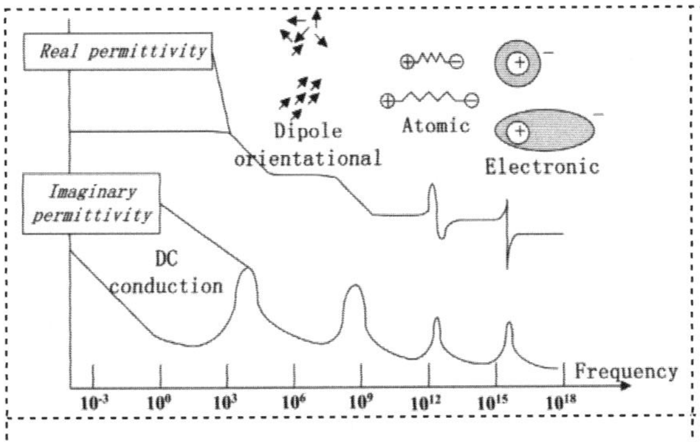

Figure 1 Dielectric permittivity spectrum over a wide range of frequencies

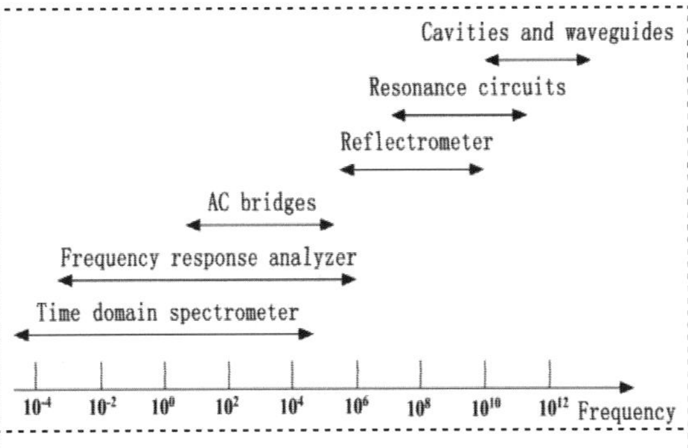

Figure 2 Summary of dielectric spectroscopy techniques

2. Choice of techniques for XLPE cables

Various dielectric spectroscopy techniques and their associated frequency ranges are summarized in Figure 2 [4]. As we are interested in the dielectric loss mechanisms in high voltage power cables operated at 50Hz, the low frequency techniques covering 10^{-4}Hz~10^{6}Hz have been explored in search of those with appropriate sensitivity to measure the extremely low loss model cables. In practice, most techniques only have sufficient sensitivity over limited frequency ranges and it is necessary to combine many techniques to achieve the required sensitivity over a wide range of frequency.

The frequency response analyzer (FRA) in conjunction with a dielectric interface has a wide frequency range. As shown in Figure 3, phase sensitive voltmeters are used to compare the voltage $U_1(\omega)$ and $U_2(\omega)$ (the latter proportional to $I_s(\omega)$). Commercial instruments often employ a reference impedance Z_R within the dielectric interface to facilitate accurate measurements. The complex impedance $Z_S(\omega)$ of the sample can be calculated from the measured data by the equation

$$Z_S(\omega) = \frac{U_S(\omega)}{I_S(\omega)} = R(\frac{U_1(\omega)}{U_2(\omega)} - 1).$$

Where ω is the angular frequency. The complex relative permittivity can be then calculated from

$$\varepsilon^* = \frac{1}{i\omega Z_S(\omega) C_0},$$

provided that the geometric capacitance, C_0, of the sample is known [5]. However, tests have shown that sensitivity of loss tangent is limited to the range 10^{-4} to 10^{-2} dependent on frequency as can be seen from the background measurements in figure 6. In practice, a Solartron FRA 1255 and Dielectric Interface 1296 were used for the measurement of the model XLPE cables.

Transformer ratio bridges (TRB), as shown in Figure 4, are often used to cover the audio frequency range. Using a set of standard impedances and a multi-tap transformer, this type of bridge allows for high precision measurements compared with other measurement techniques. However, the frequency range is more restricted than the FRA. When the currents I_1 and I_2 flowing through standard Z_S and unknown Z_U are equal in magnitude,

$$\frac{V_1}{Z_S} = \frac{V_2}{Z_U}$$

they will combine to produce zero core flux in the current transformer. Zero current in the neutral line can be detected with a null indicator, because the two currents will be 180° out of phase to each other. In this study, Wayne Kerr universal bridge B221 was used to determine the dielectric loss for XLPE cable systems over the frequency range of 200Hz~20kHz and loss tangent resolution of below 10^{-4}.

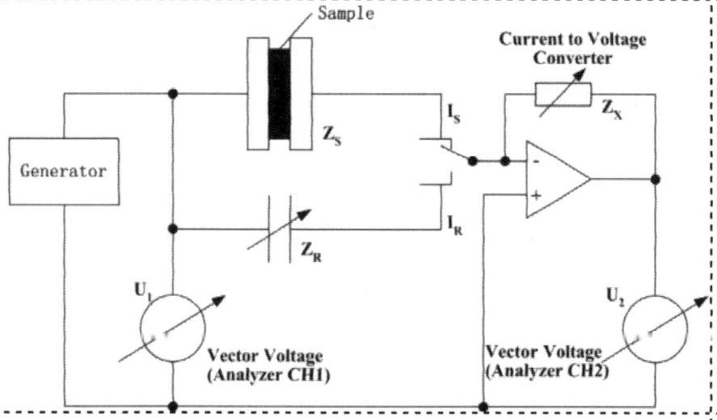

Figure 3 Principle of frequency response analyzer

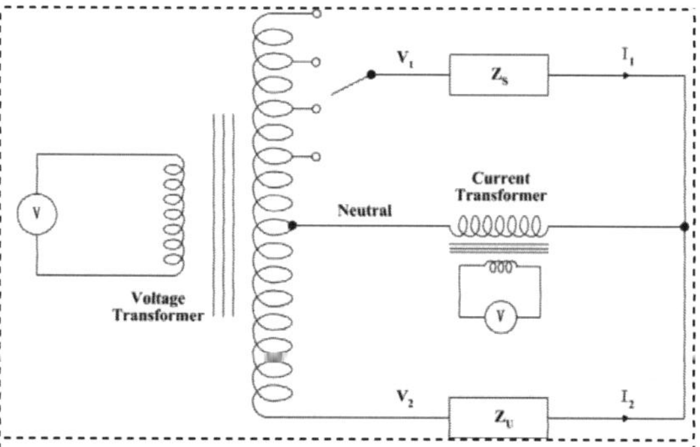

Figure 4 Principle of transformer ratio bridge

In order to achieve sufficient sensitivity around 50Hz, alternative methods were employed involving time domain dielectric spectroscopy. This involves time domain measurement of the charging/discharging currents. The principle of obtaining frequency spectra from current measurement is shown in Figure 5. In this study, an electrometer was used to measure the discharging current of XLPE cable samples, I(t), with the time domain data captured using a digitizing oscilloscope. The complex permittivity was determined using the Fourier transform,

$$\varepsilon^*(\omega) = \varepsilon_\infty + \frac{1}{C_0 V_0}\int_0^\infty I(t)\exp(-j\omega t)dt - j\frac{G}{\omega C_0},$$

where ε_∞ is the high frequency permittivity, V_0 is the applied dc voltage and G is the conductance.

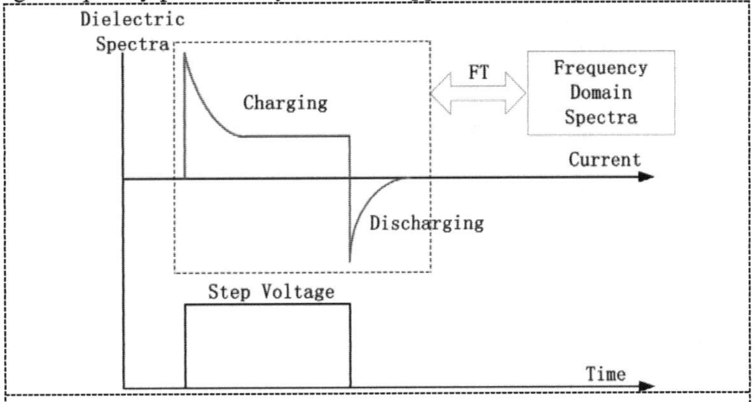

Figure 5 Principle of time domain dielectric spectroscopy

3. Experiment results

The XLPE model cables were produced by Borealis AB and degassed for 5 days at 80C, before they were prepared into 5m long test samples. Conductive adhesive copper tape was used to wrap the cable sample in order to make good electrical contact with the outer semicon layer.

3.1. *Results of different techniques*

FRA measurement results are shown in Figure 6 as a function of cable temperature. DC conduction behaviour (slope of -1 in the tanδ response) can be observed above 40C in the range of 10^{-4}Hz~1Hz. At frequencies above 10Hz, the results cannot be distinguished from the instrument background; measured using a specially designed empty cell cable. This was made of copper with the same geometry as the model cables. Arrhenius behaviour, with an activation energy of 1.02eV, was found in the DC conductivity using the equation

$$\sigma = \sigma_0 \times \exp(\frac{E}{kT}).$$

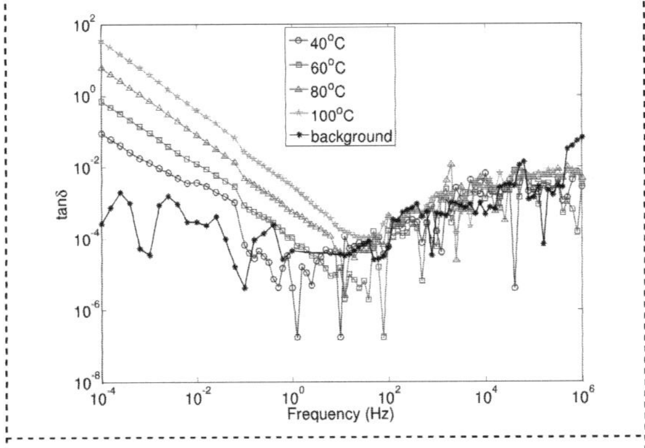

Figure 6 FRA results on XLPE cables

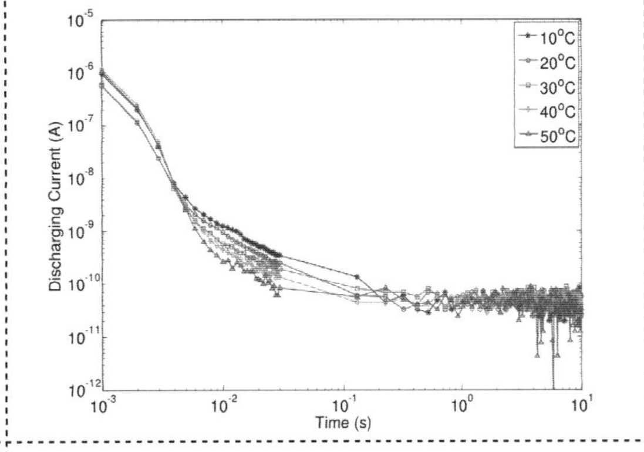

Figure 7 Discharging current of XLPE cables

TDDS (time domain dielectric spectroscopy) measurements were obtained from discharging current measurements taken over the range 1ms~10s (as shown in Figure 7). Excluding unreliable data due to truncation, noise and the DFT algorithm, the usable spectroscopy range was about 0.5~200Hz.

The frequency spectra were calculated using equation

$$\varepsilon^*(k) = \varepsilon_\infty + \frac{2}{NC_0V}\sum_{i=1}^{N} x(i)\exp\left\{\frac{-2\pi j}{N}ik\right\} \quad k = 0,1,2,\ldots N$$

and are shown in Figure **8**. The loss tangent of the XLPE cables was found to be around 10^{-4} at about 1Hz and increases to 10^{-3} with a peak at approximately 200Hz.

Transformer bridge measurement results with different with and without the outer copper tape are shown in Figure 9. Additional losses above 3kHz are associated with loss due to the conductance of the semicon material. Hence for frequencies above this value, copper conductive tape must be used to improve electrical contact along the whole length of the cable. With copper tape, the loss tangent of XLPE cables in this frequency region decreases from 10^{-3} to below 10^{-4} at 6kHz.

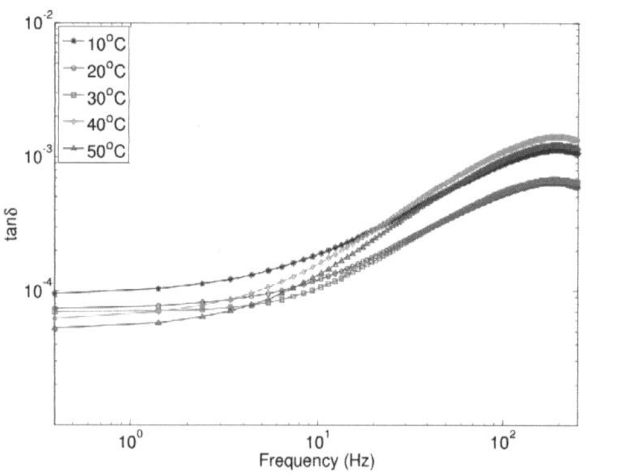

Figure 8 Frequency spectra of TDDS data after FT

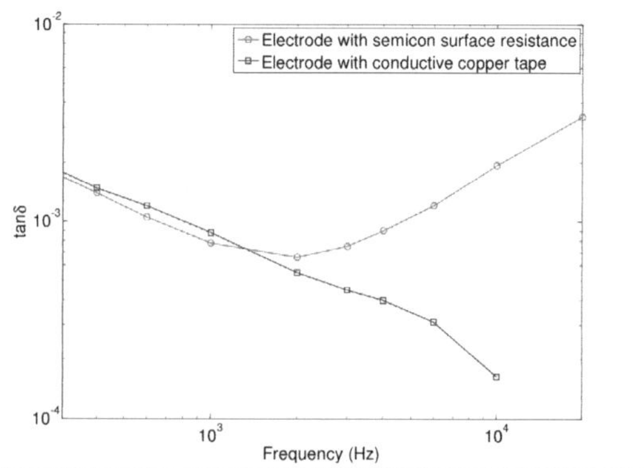

Figure 9 TRB results on XLPE cables with different measuring electrodes

3.2. Master curve of XLPE cable spectra

The 40C measurement data using the 3 spectroscopy techniques was merged together to give the master curve of the XLPE cables in a frequency range from 10^{-4}Hz to 10^{4}Hz, as shown in Figure 10.

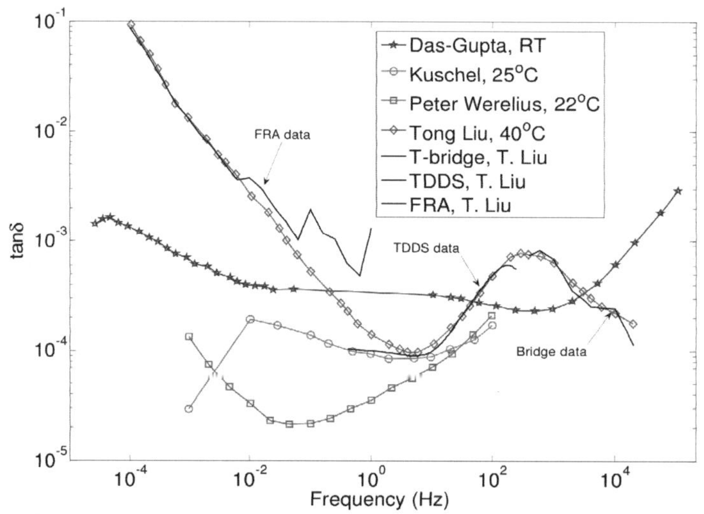

Figure 10 Comparison of XLPE cable spectra with Das-Gupta, Kuschel, and Peter Werelius

At low frequencies, the master curve exhibits DC conduction and the conductivity of the XLPE cable at 40C was calculated to be 1.1×10^{-15}S/m. A loss peak was also found at 200Hz at the join of the TDDS data and the bridge data. In Figure 10, published data by Das-Gupta [6], Kuschel [7] and Peter Werelius [8] [9] are shown for comparison. Using high voltage dielectric spectroscopy using a frequency response analyzer, Peter Werelius measured 12kV XLPE cables from 10^{-3}Hz to 100Hz. At lower frequencies, a DC conduction can be identified with the significantly smaller conductance (compared with the master curve) being due to the lower test temperature. Kuschel also measured non-aged 12/20kV XLPE cables of different German manufacturers with an active length of between 0.5m and 2m in Technid University of Berlin. Very low flat loss was found in the loss tangent spectra at 1kV and 25C. The dielectric loss is also increasing at higher frequency which is similar with the master curve in this study. Combining both FDDS and TDDS techniques, Das-Gupta investigated a wide frequency range loss spectra on 15kV XLPE cable samples. The lower frequency spectrum was transformed from discharging current with Hamon Approximation. While no DC conduction should be included in the TDDS results, a low frequency peak was assumed by fitting the data with the Universal Relaxation Law [10]. The middle frequency range is flat without any measurement data and another broad peak was fitted by Das-Gupta. The additional loss at higher frequencies was due to the conductance of tap water, used as an outer measuring electrode.

4. Conclusion and future work

In conclusion, the measurement results have identified 3 dielectric loss mechanisms:

1. In the lower frequency range, below 1Hz, DC conduction behaviour was found to be dominant.
2. In the middle frequency range of 100Hz~300Hz, a relaxation peak was found centred at a frequency of 200Hz.
3. In the higher frequency range of 500Hz~20kHz, the electrical conductance of the outer semicon layer contributes additional dielectric loss. This can be eliminated by improved electrical contact to the outer semicon material using copper tape.

Future work includes improving the time domain technique and validating the loss peak at 200Hz. The loss mechanisms of XLPE cables will be further studied with using different types of insulation/semicon materials as well as non-degassed and aged cables.

5. References

[1] R. M. Black, *The History of Electric Wires and Cables*, Peter Pergrinus, London 1983.
[2] Vahdat Vahedy, "Polymer Insulated High Voltage Cables", IEEE Electrical Insulation Magazine, Vol. 22, No. 3, 2006.
[3] Walter S. Zaengl, "Applications of Dielectric Spectroscopy in Time and Frequency Domain for HV Power Equipment", IEEE Electrical Insulation Magazine, Vol. 19, No. 6, 2003.
[4] Tony Blythe, David Bloor, *Electrical Properties of Polymers*, Cambridge University Press, 2005.
[5] James P. Runt, John J. Fitzgerald, *Dielectric Spectroscopy of Polymeric Materials: Fundamentals and Applications*, American Chemical Society, 1999.
[6] P. C. N. Scarpa, A. Svatík, D. K. Das-Gupta, "Dielectric spectroscopy of polyethylene in the frequency range of 10^{-5} Hz to 10^{6} Hz", Polymer Engineering and Science, Vol. 26, No. 8, 1996.
[7] M.Kuschel and W.Kalkner, "Dielectric response measurements in time and frequency domain ofdifferent XLPE homo- and copolymer insulated medium voltage cables", IEE Proceedings - Science, Measurement and Technology, Vol. 146, No. 5, 1999.
[8] Peter Werelius, *Development and Application of High Voltage Dielectric Spectroscopy for Diagnosis of Medium Voltage XLPE Cables*. PhD thesis, Stockholm: KTH, Electrical Engineering, 2001.
[9] P. Werelius, P. Tharning, R. Eriksson, B. Holmgren, and U. Gafvert, "Dielectric spectroscopy for diagnosis of water tree deterioration in XLPE cables", IEEE Transactions on Dielectrics and Electrical Insulation, Vol. 8, No. 1, pp. 27–42, 2001.
[10]A. K. Jonscher, *Universal Relaxation Law*, Chelsea Dielectrics Press, London, 1996

Acknowledgements
The author is grateful to Mr Ulf Nilsson and Borealis AB Sweden for the sponsorship.

Journal of Physics: Conference Series **183** (2009) 012003 doi:10.1088/1742-6596/183/1/012003

High signal to noise ratio THz spectroscopy with ASOPS and signal processing schemes for mapping and controlling molecular and bulk relaxation processes

S. Hadjiloucas[1], G.C. Walker[1], J.W. Bowen[1], V.M. Becerra[1], A. Zafiropoulos[2] and R.K.H. Galvão[3]

[1]Cybernetics, School of Systems Engineering, The University of Reading, RG6 6AY, UK
[2]Biosystems Engineering Department, School of Agricultural Technology, Technological Educational Institute of Larissa, 411 10, Larissa, Greece.
[3]Divisão de Engenharia Eletrônica, Instituto Tecnológico de Aeronáutica, São José dos Campos, SP, 12228-900 Brazil.

Correspondance email: s.hadjiloucas@reading.ac.uk

Abstract. Asynchronous Optical Sampling has the potential to improve signal to noise ratio in THz transient sperctrometry. The design of an inexpensive control scheme for synchronising two femtosecond pulse frequency comb generators at an offset frequency of 20 kHz is discussed. The suitability of a range of signal processing schemes adopted from the Systems Identification and Control Theory community for further processing recorded THz transients in the time and frequency domain are outlined. Finally, possibilities for femtosecond pulse shaping using genetic algorithms are mentioned.

1. Introduction

After discussions with A. Bartels of GigaOptics GmbH at an NPL meeting in 2005, where S. Hadjiloucas of Reading University proposed an alternative frequency comb interrogation concept, a dual resonator 35 femtosecond pulse risetime Ti:sapphire prototype spectrometer was developed in Germany to perform THz transient time domain spectrometry (TDS) using an Asynchronous Optical Sampling Scheme (ASOPS) [1-3]. The final system delivered at Reading incorporated large surface area Auston switches with inter-digitated electrodes (TeraSEDs) which were conceptualized in Germany [4], to generate THz pulses from mid-infrared (800 nm) pulses. This scheme can provide an improvement in signal to noise ratio by a factor of 20 when compared to conventional THz transient spectrometers that use the traditional scanning delay line to perform the cross-correlation between pump and probe beams.

An important advantage of eliminating the translation stage from the spectrometer is that the spot size of the optical beam propagating through the reference path of the interferometer is no longer of variable size (due to diffractive spreading of the infrared beam) for different path lengths imposed by the translation stage, as is in the case for a conventional THz transient spectrometer. This is a very important advantage in ASOPS as a dilution of the gate pulse in an Auston receiver changes its multi-moded antenna pattern in both amplitude and phase delay. This type of error has not been systematically addressed by the THz community although careful work [5] has shown that changing the spot size dramatically alters the recorded time-domain signature and hence the corresponding Fourier transformed spectrum. Such errors are endemic to most THz transient spectrometers and can be exacerbated when ratioing the background and sample spectra.

In addition, although not extensively discussed by the THz community, the THz focusing elements in the system (paraboloids) have also a detrimental effect in the way the radiation propagates through the sample. The beam is diffractively spreading and can no longer be considered as a plane wave incident normally upon the sample. As a consequence, there are cases where not all of the radiation reflected from or transmitted through the sample is detected. One way of gaining further understanding of these phenomena in a quantitative way is to create a full vectorial description of the THz beam incident on the sample and investigate the effect of different degrees of focusing to the observed spectra in a systematic manner. Using information from the focusing elements in the system, the diffractively spreading THz beam can then be represented as a weighted distribution of plane waves incident on the sample over a range of possible angles of incidence [5]. The modification of these plane waves in amplitude, phase and polarisation content should then be propagated through the sample.

Finally, the detection techniques employed in THz TDS are spatially and polarization selective and any sample-induced distortion of the spatial profile and polarization of the THz beam will modify the detected signal in a manner which may mimic absorption and phase delay in the sample [6]. Similar errors are commonly encountered in electro-optic detection schemes, although most of the THz community seems to ignore them.

In the following section, a new inexpensive control loop is suggested for controlling the slave repetition rate laser in ASOPS. In section three, some useful signal processing schemes for THz transients developed over the years are outlined. In section four, a new femtosecond pulse shaping methodology is mentioned. General concluding remarks and applications to these developments are mentioned.

2. Frequency to voltage converter based ASOPS control scheme

Figure 1 depicts the two Verdi-pumped femtosecond Ti:sapphire ring oscillators contained in the GigaJet 20, one of them (slave oscillator) has a mirror mounted on a piezoelectric transducer to control its frequency comb repetition rate at an offset of 20 kHz from the other (master) oscillator.

A schematic of the developed ASOPS THz set-up showing the control loop for the synchronization of the repetition rate of the two femtosecond ring resonators is shown in figure 1. The two femtosecond lasers pumped by two 5 W Verdi lasers produce femtosecond pulses with a repetition rate f and $f+\Delta f$ set by the optical path length of each resonator (around 1 GHz when free running). Using beam splitters, a small portion of the signal (ratio 90:10) is directed to fast photodiodes which produce electrical trigger signals which are further amplified by low-noise high-bandwidth microwave amplifiers (model ZFL 1000LN from Mini-circuits) operated in a trans-impedance configuration (converting a photo-current into a voltage). Although the photodetectors are not preserving the shape of the pulse, this is of no consequence to the stabilization scheme as the goal is to generate a triggering event for the synchronization process. The difference in repetition rate ($\Delta f \approx 20$ kHz) is obtained using a microwave mixer and sent to an AD650 (Analog Devices) frequency to voltage converter (100 kHz full scale with 0.005% linearity) which provides an output between 0 and 10 V. This output is subsequently fed to a SIM 960 analog PID controller (Stanford Research Systems) before being finally sent to a pair of PA240CX (APEX semiconductor) op-amps, connected in a bridge configuration, providing 660 V peak to peak (circuit design implementation as found in Apex Semiconductors Application note 25) to drive the piezoelectric transducer that controls the path length in the slave resonator cavity. The bandwidth of the high voltage amplifier and piezo combination was in excess of 100 kHz ensuring elimination of acoustical noise in the system. For a 30 cm resonator length, the round trip time for a pulse in the resonator is 1 nanosecond.

The designed close-loop system has a large linear gain in the forward path (composed of the cascaded PID signal, high voltage amplifier signal, the transducer voltage to displacement conversion, and the conversion of resonator path length to frequency comb repetition rate) and unity gain in the feedback path. With reference to figure 1, using transfer function notation and using $A >> 1$:

$$V_{out} = \frac{A}{1+AB} V_{REF} \approx \frac{A}{AB} V_{REF} = \frac{V_{REF}}{B} \tag{1}$$

so that the characteristics of the system are governed by the stable characteristics of the feedback path B and the difference in the repetition rate between the two resonators is set by V_{REF}. Controlling the path length in one resonator to one part per thousand (i.e., to a precision of 300 micrometers) using the piezo, therefore, means that a timing jitter of approximately 1 picosecond can be easily achieved with this configuration. This jitter is due to the noise signal at the output of the frequency to voltage converter (which is currently around 10 mV). This can be improved upon with better shielding of the electronics.

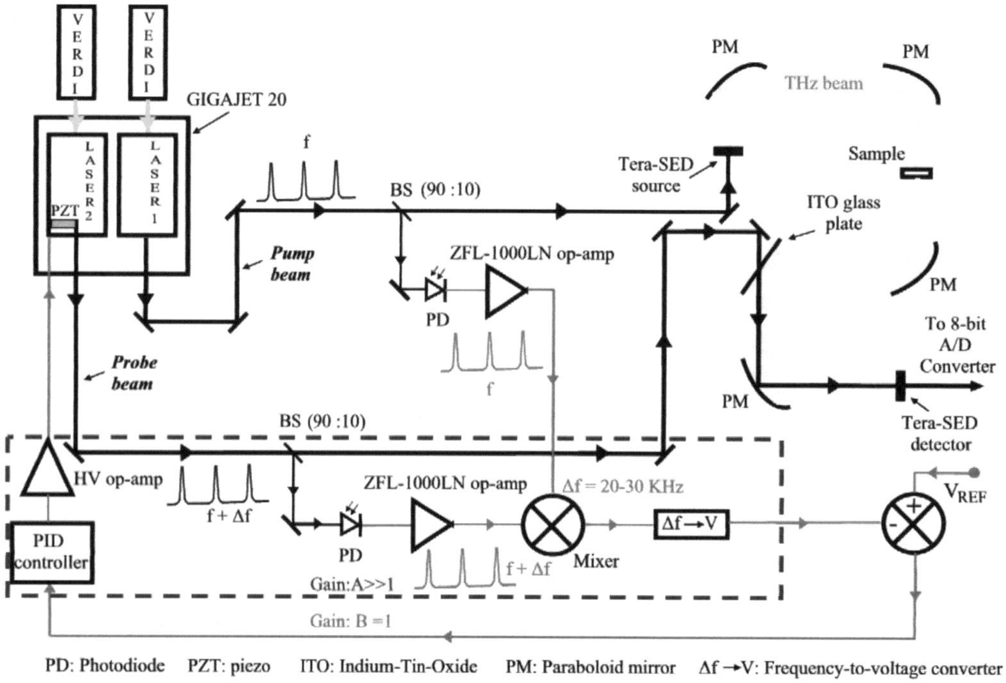

Figure 1. ASOPS THz transient prototype spectrometer based on the GigaOptics Gigajet 20 system and feedback locking scheme.

In figure 1, the overall master-slave repetition rate control scheme is shown. The triggering of the TeraSED detector is performed by the optical pulses from the slave laser. Although two Tera-SEDs are used one for generating and one for detecting the THz transients, it is also possible to use an electro-optic detection set-up (ZnTe crystal and Wollaston prism) with a balanced photodetection scheme to sample the THz transient. The existing 8-bit analog to digital converter used for data acquisition in the current experiments (which is at the front end of the 50 GHz Agilent Infinium oscilloscope we currently use) is not the optimal choice for digitising the observed transients and a dedicated 12 bit or 14 bit data acquisition card with a bandwidth of 100 MHz would be a more appropriate solution to implement, before we can report on the capabilities of the overall system.

3. Signal processing schemes of femtosecond duration transients

In the area of signal processing (de-noising of signals) generated from THz transient spectrometers, we have developed a range of useful algorithms [7,8] to de-embed the complex propagation constant from multi-modal excitation of micro-machined waveguides (useful for micro-spectroscopy applications where the waveguide is filled with small volumes of the dielectric to be characterised). This is achieved by adopting algorithms found in the Systems Identification literature (ARX, and N4SID

Subspace models for linear spectroscopy and NARX models to account for the non-linear cut-on of waveguide modes). Since a femtosecond duration pulse is capable of persistent excitation of the medium within which it propagates, such approach is perfectly justifiable. The multimode Brownian oscillator time-domain response function described by state-space models is a mathematically robust framework that can be used to describe the dispersive phenomena governed by Lorentzian, Debye and Drude responses. In addition, the optical properties of an arbitrary medium can be expressed as a linear combination of simple multimode Brownian oscillator functions.

Additional routines developed at Reading include wavelet filtering [9] of THz transients (using conventional mother wavelets and adaptive structures at each decomposition level), PCA classification schemes [10] and novel apodization structures [11] as the time-domain background and sample interferograms are non-symmetrical. These procedures have useful de-noising properties and can lead to a more precise estimation of the complex insertion loss function.

Current work concentrates on the development of novel system identification tools in the wavelet domain for the further processing of THz transients. Defining the background and sample interferograms as the input and output signals, the frequency response of an identified model would be an estimate of the complex insertion loss (CIL). A wavelet-packet formulation illustrated in figure 2 is adopted and sub-band models $\mathbf{M}_{i,j}(z)$ are identified from the sample and background interferograms by following a least-squares procedure as indicated in figure 3.

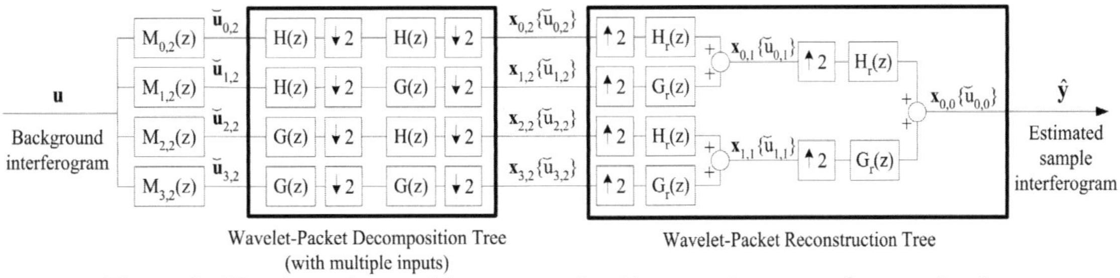

Figure 2. Wavelet-packet model structure. In this example, a complete two-level decomposition tree, which defines four frequency sub-bands, is employed. H(z), G(z) denote low-pass and high-pass decomposition filters, respectively, with reconstruction counterparts represented by $H_r(z)$, $G_r(z)$. The four sub-band models are represented by the transfer functions $\mathbf{M}_{0,2}(z)$, $\mathbf{M}_{1,2}(z)$, $\mathbf{M}_{2,2}(z)$, $\mathbf{M}_{3,2}(z)$.

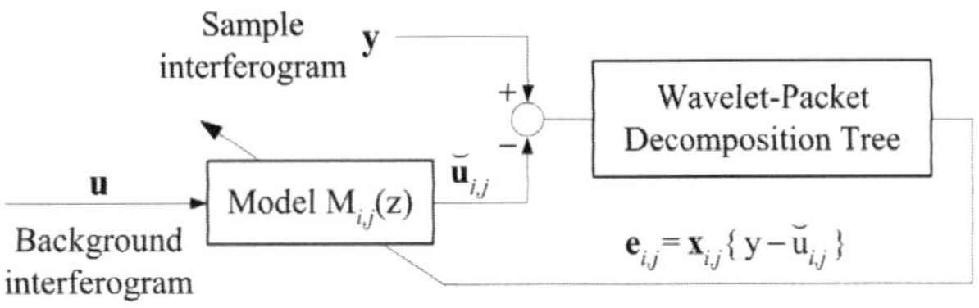

Figure 3. Model identification of a sample interferogram for a given frequency sub-band.

Figure 3 illustrates the procedure adopted to identify each sub-band model $\mathbf{M}_{i,j}$. $\mathbf{u}$ is the input signal used for identification. $\mathbf{y}$ and $\breve{\mathbf{u}}_{i,j}$ are the plant and sub-band model outputs, respectively. Residue $\mathbf{e}_{i,j}$ denotes the wavelet-packet coefficients of the difference between y and $\breve{\mathbf{u}}_{i,j}$, in the frequency band under consideration. The structure adopted for the sub-band model is a transfer function of the form: $\mathbf{M}_{i,j}(z) = \mathbf{P}_{i,j}(z)\mathbf{Q}_{i,j}(z)$ where $\mathbf{P}_{i,j}(z) = \left(1 - z^{-1}\right)^{-s_{i,j}}$, $s_{i,j} \in Z$ and $\mathbf{Q}_{i,j}(z) = \alpha_{i,j} + \beta_{i,j} z^{-1}$, with $\alpha_{i,j}, \beta_{i,j} \in \Re$. $\mathbf{P}_{i,j}(z)$ is aimed at roughly approximating the band-limited frequency response of the plant, whereas the Finite Impulse Response (FIR) term $\mathbf{Q}_{i,j}(z)$ provides a fine-tuning for the approximation. A least-squares adjustment for the parameters of $\mathbf{M}_{i,j}$ can be carried out by minimizing the following cost function $\mathbf{J}_{i,j}: Z\times\Re^2 \rightarrow \Re$: $\mathbf{J}_{i,j}(s_{i,j},\alpha_{i,j},\beta_{i,j}) = \mathbf{e}_{i,j}(\mathbf{e}_{i,j})^{\mathrm{T}}$ where $\mathbf{e}_{i,j}$ denotes the row vector of residues for the identification data, as shown in Fig. 3.

4. Genetic algorithms developments for femtosecond pulse shaping experiments

Another line of work at Reading involves the development of a series of evolutionary meta-algorithms in MATLAB, running within the LABVIEW environment to perform pulse shaping of femtosecond duration laser pulses. Proof of principle experiments were performed where the signal from a LABVIEW-controlled function generator was sent to an RF-excited acousto-optic modulator crystal which controls the spectral content and time domain characteristics of 120 femtosecond duration pulses generated from an optically pumped Ti:sapphire Kerr lens mode-locked laser oscillator. The pulses have a useful Fourier transform limited bandwidth of 50-80 nm around a center frequency of 800 nm. The genetic algorithm (GA) population in the developed software consists of a collection of waveforms, each made from two concatenated vectors, one specifying phases over a range of frequencies and the other specifying corresponding magnitudes. Amplitude and phase genes are used to search the evolutionary landscape for desired pulse shapes with specific characteristics. With appropriate pulse shaping schemes realised in pump-probe experiments, it should in principle allow us to control bulk relaxation processes at picosecond timescales.

Selecting a breeding subset from the N best fit members (amplitudes and phases) is a very greedy heuristic that often leads to convergence around local optima instead of the global solution. One attempt to circumvent this problem is to use a probabilistic selection of breeding candidates. For instance, with roulette wheel selection (RWS), each member of the population is assigned a slice with an area proportional to fitness. Then, using the proportional probabilities from the roulette wheel, one or two candidates are selected out of the entire population and passed to the operators for production of children. The process is repeated again and again until the desired number of children are created. Because members are never removed from the roulette wheel, members with high fitness are virile and, consequently, they are chosen as parents more frequently than those with low fitness. The process can easily be coded into the algorithm by firstly creating a vector F of corresponding fitnesses for the population and then creating a cumulative summed vector for each of all indices in F.

As the population starts to converge to a solution, the average fitness of the population will also start to converge to the best fitness. In RWS, the slice size for each population member will be roughly the same size, leaving little differentiation between worst and best fit members. This in turn has a detrimental effect on the selection process, and slows down the convergence of the algorithm considerably. To keep the selection pressure high, we employ fitness scaling and use the scaled fitnesses in RWS instead. This is performed by assigning the best fit member a scaled fitness of 2 and the average fitness a scaled fitness of 1. This way, the best fit member is twice as likely to be chosen as the average fit member.

After measuring and then linearly scaling the fitness, an operator is selected by RWS from a vector of operator weights. Depending on the operator chosen, either one or two parents are selected and then passed on to the operator, thereby creating the child. Initially, the algorithm used seven different operators. Six of these directly exploited the frequency domain: two-point crossover, average crossover, mutation, polynomial phase mutation, creep, and three point smoothing. Current work concentrates in developing a differential evolution scheme and operators in the wavelet domain to perform high resolution waveform matching with a much higher convergence rate.

5. Conclusion

A simple inexpensive control scheme for adjusting the repetition rates of two Ti:sapphire laser oscillators at an offset frequency of 20 kHz has been proposed. The scheme is appropriate for asynchronous optical sampling experiments in THz transient spectrometry, but due to $1/f$ noise and the delay of the frequency to voltage converter in translating the low frequencies into a voltage, it would be less suitable in applications where the offset between the two oscillators must remain at 1 Hz or less [12]. Performance is limited by the linearity of the frequency to voltage chip used in the implementation.

The signal processing of time domain THz transients has been discussed in general terms and a new systems identification approach which can be implemented in the wavelet domain is proposed. The schemes are particularly promising for de-noising purposes as well as in chemometric applications where mixtures of absorbing species are studied.

Finally, a femtosecond pulse shaping methodology is suggested for the optical pulses generated by the femtosecond lasers. The methodology is appropriate for adaptive equalization applications where the goal would be to compensate for dispersion in optical systems such as fibre-optic networks, in high harmonic generation crystals and optical parametric oscillators. Femtosecond pulse shaping techniques in the weak and strong field regimes have the potential to perform mapping of potential energy surfaces in molecular reactions as well as elucidate ultrafast energy transfer dynamics in chemical reactions [13]. Light energy can act as a catalyst to facilitate the progress of molecular and bulk relaxation processes and as a consequence, control the progress of reactions, enabling us to gain further insights in reaction kinetics. Using THz type-setting techniques, there is also the potential to modulate THz transients for very high bit rate communications over short distances.

6. References

[1] C. Janke, M. Först, M. Nagel, H. Kurtz, and A. Bartels, 2005. 'Asynchronous optical sampling for high-speed characterization of integrated resonant terahertz sensors', *Opt. Lett.,* **30**, 1405-1407.

[2] A. Bartels, A. Thoma, C. Janke, T. Dekorsy, A. Dreyhaupt, S. Winnerl, and M. Helm, "High-resolution THz spectrometer with kHz scan rates," *Opt. Express* **14**, 430-437 (2006)

[3] A. Bartels, R. Cerna, C. Kistner, A. Thoma, F. Hudert, C. Janke, and T. Dekorsy (2007) 'Ultrafast time-domain spectroscopy based on high-speed asynchronous optical sampling' *Rev. Sci. Instrum.* **78**, 035107

[4] A. Dreyhaupt, S. Winnerl, T. Dekorsy, and H. Helm, (2005) "High-intensity terahertz radiation from a microstructured large-area photoconductor," *Appl. Phys. Lett.* **86,** 121114-121111.

[5] G.C. Walker 2003 '*Modelling the Propagation of Terahertz Radiation in Biological Tissue*' PhD Thesis, Centre of Medical Imaging Research, University of Leeds.

[6] J.W. Bowen, G.C. Walker and S. Hadjiloucas 2007 'Sample-Induced Beam Distortions in Terahertz Time Domain Spectroscopy and Imaging Systems', *Joint 32nd International Conference on Infrared and Millimetre Waves and 15th International Conference on Terahertz Electronics,* Cardiff, UK 208-209.

[7] S. Hadjiloucas, R. K. H. Galvão, V. M. Becerra, J. W. Bowen, R. Martini, M. Brucherseifer, H. P. M. Pellemans, P. Haring Bolívar, H. Kurz, J. M. Chamberlain, 2004 'Comparison of state space and ARX models of a waveguide's THz transient response after optimal wavelet filtering,' *IEEE Transactions on Microwave Theory and Techniques*, **52**, 2409-2419.

[8] R.K.H. Galvão, S. Hadjiloucas, V.M. Becerra and J.W. Bowen, 2005 'Subspace system identification framework for the analysis of multimoded propagation of THz-transient signals,' *Measurement Science and Technolgy*, **16**, 1037-1053.

[9] R.K.H. Galvão, S. Hadjiloucas, J.W. Bowen and C.J. Coelho, 2003 'Optimal discrimination and classification of THz spectra in the wavelet domain,' *Optics Express*, **11**, 1462-1473.

[10] S. Hadjiloucas, R.K.H. Galvão and J.W. Bowen, 2002 'Analysis of spectroscopic measurements of leaf water content at THz frequencies using linear transforms,' *Journal of the Optical Society of America A* **19**, 2495-2509,.

[11] R.K.H. Galvão, S. Hadjiloucas, A. Zafiropoulos, G.C. Walker, J.W. Bowen and R. Dudley 2007 'Optimization of Apodization Functions in THz Transient Spectrometry,' *Opt. Lett.,* **32**, 3008-3010.

[12] F. Keilmann, C. Gohle, and R. Holzwarth, (2004). 'Time-domain mid-infrared frequency-comb spectrometer', *Opt. Lett.*, **29**, 1542-1544.

[13] J. Savolainen, N. Dijkhuizen, R. Fanciulli, P.A. Liddell, D. Gust, T.A. Moore, A.L. Moore, J. Hauer, T. Buckup, M. Motzkus, and J.L. Herek (2008) 'Ultrafast Energy Transfer Dynamics of a Bioinspired Dyad Molecule' *J. Phys. Chem. B*, **112** (9), 2678–2685

Space charge suppression effect of nano-size fillers added to polymeric materials

Y Hayase[1], Y Tanaka[1], T Takada[1], Y Murata[2],Y Sekiguchi[2] and C C Reddy[2]

1.Tokyo City University, 1-28-1, Tamatsutsumi, Setagayaku, Tokyo, JAPAN
2.J-Power Systems Corp., 5-1-1, Hitaka, Hitachi, Ibaraki, JAPAN

E-mail: tanaka@eml.mes.musashi-tech.ac.jp

Abstract. Space charge suppression mechanism in nano-composite polymer material is studied using experimental results and numerical simulation. Recently, many kinds of nano-composite polymeric materials have been reported to have improved their characteristics under high electric field. For example, LDPE/MgO nano-composite, which is made up of low density polyethylene (LDPE) and nano size filler of magnesium oxide (MgO), exhibits high volume resistivity and high dielectric strength under dc electric field. Authors have investigated the space charge behaviour in LDPE/MgO nano-composite under high electric field using pulsed electro-acoustic (PEA) method. It has been found that, compared to LDPE, the space charge formation is also suppressed in the nano-composite material. As a reason for the suppression, we have suggested that the induced dipole polarization around MgO filler formed by dc stress application might play a role of carrier trap sites. From the numerical calculation, distortion of electric potential around MgO is seen to be much larger than that around naturally included dipole. It means that the MgO acts as a deep trap site as different from some defect or ions included in LDPE. Using the numerical calculation based on such electric potential distortion, we have tried to simulate the space charge distribution in LDPE/MgO under high dc electric field. The simulation results are in good agreement with the experimental results.

1. Introduction

Polymer nano-composites have drawn researchers' attention as next generation insulating materials [1-3]. Addition of small amount only of a nano sized filler to a base polymer, electrical properties of the composite material are drastically improved. For example, LDPE/MgO nano-composite, which is made up of low density polyethylene (LDPE) and nano size filler of magnesium oxide (MgO), shows high resistivity and dielectric strength under dc electric field [4, 5]. Furthermore, authors have found that space charge accumulation is also drastically suppressed under high dc electric field in LDPE/MgO nano-composite. Authors have been investigating the space charge behavior in LDPE under very high dc electric field of more than 100 kV/mm using the pulsed electro-acoustic (PEA) method [6]. Under such stress, what is called a 'packet-like charge' [7, 8], is injected from an electrode and it distorts electric field distribution in bulk of LDPE significantly. In some cases, the distorted electric field distribution has a maximum value of twice the applied average electric field, and often a breakdown occurs in the sample [8]. Earlier, authors have given some explanation for such packet-like charge behavior numerically [9]. However, under same test conditions as that of LDPE, the packet-like charge doesn't appear in the LDPE/MgO nano-composite [10]. It means that the nano-composite material is very reliable material for use under high dc electric field. However, the mechanism of the suppression has not been clear yet. Considering the improvement in electrical properties, Prof. Tanaka et.al. [11] have proposed that, a "multi-core" zone induced on the surface of nano filler may play a role of such drastic change. However, by only such explanation, it is hard to understand the suppression of space charge accumulation in LDPE/MgO under high dc stress. Therefore, authors have tried to explain it in terms of the induced dipole around additional filler in polymers under dc electric field [12]. In this paper, authors try to show the comparison between numerical simulations based on such idea and typical measurement results.

2. Space charge accumulation in nanocomposite films

2.1. Space charge accumulation in LDPE and LDPE/MgO under high dc stress.

At first, the difference between space charge accumulations in LDPE and LDPE/MgO nano-composite is briefly introduced. All samples are supplied from J-power systems Corp. Figure 1 shows typical space charge measurement results using PEA system under dc high electric fields of 100, 150 and 200 kV/mm. In this figure, time dependent space charge profiles are described using color chart. In the color chart, the red and blue colors stand for the positive and the negative charge densities, respectively. The color bar described beside the data shows the scale of charge density for color chart. In this figure, the vertical axes stands for the time after voltage application in minutes. The horizontal axis stands for the position in the bulk of sample. Left (blue) and right (red) vertical lines show the positions of induced charges on the cathode and the anode, respectively. The thickness of samples is about 70μm. In the case of LDPE/MgO nano-composites, the added amounts of MgO are 0.2 and 1phr (parts per hundred parts of resins: LDPE/MgO 1phr means 100g LDPE mixed with 1g MgO).

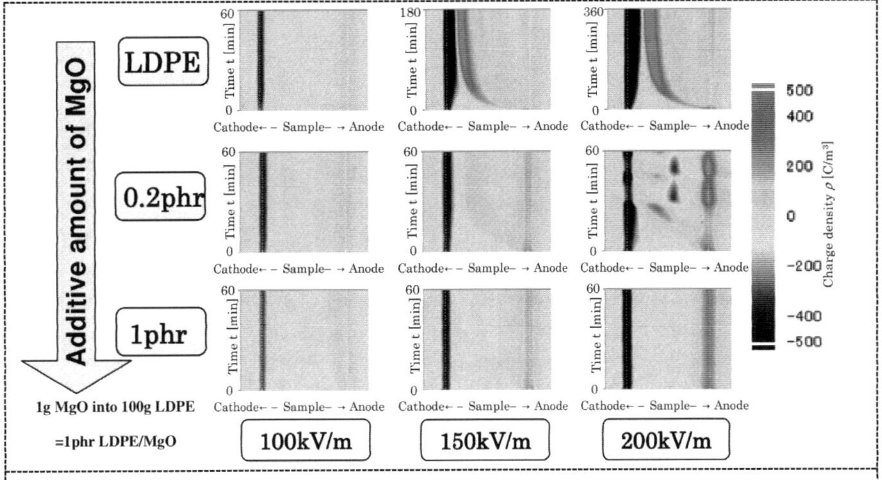

Figure 1. Time dependence of space charge distributions in LDPE and LDPE/MgO nano-composite films under dc applied electric fields of 100, 150 and 250 kV/mm.

When the average electric fields of 100 kV/mm is applied to the samples, only induced charges are observed in all samples at the interfaces between samples and electrodes. On the other hand, a large amount of "packet-like charges" is observed in LDPE under 150 and 200 kV/mm. Positive packet-like charge is injected form the anode and they move towards the cathode slowly in both cases. Since the packet-like charge gradually increases with time, it deforms the electric field distribution in bulk of the sample. The enhanced maximum electric field sometimes shows the value of twice or more of the initially applied average electric field. However, almost no such significant space charge is observed in LDPE/MgO (1 phr) even under applied average electric field of 200 kV/mm. In the case of LDPE, a breakdown sometimes occurs by the enhancement of the electric field. It means that the breakdown by such electric field enhancement hardly occurs in LDPE/MgO nano-composite. From these results, we found the addition of nano size MgO filler to LDPE shows a suppression effect for charge injection under dc stress. Details of the charge accumulation characteristics in LDPE and LDPE/MgO nano-composites are described elsewhere [10].

2.2. Space charge distributions in LDPE/MgO - Dependence on electric field

When a high electric field such as 200 kV/mm is applied to LDPE/MgO (1 phr), almost no significant space charge injection is observed. However, a small amount of space charge accumulates in bulk of LDPE/MgO (1 phr) under relatively low applied electric field. Figure 2 shows space charge distributions under applied electric fields of (a) 50 kV/mm and (b) 250 kV/mm.

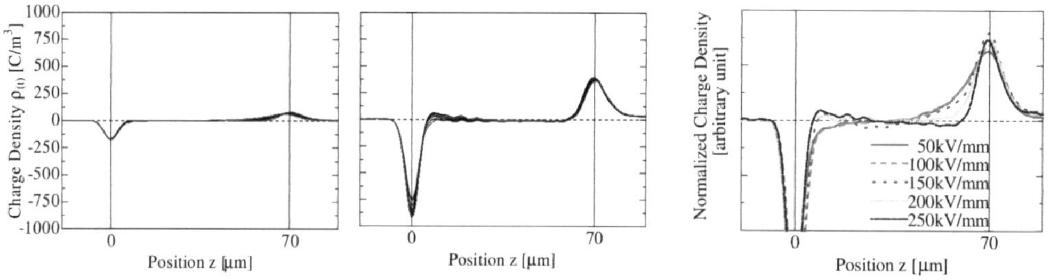

Figure 2. Space charge distributions dependence on applied electric field in LDPE/MgO (1 phr) under electric fields from 50 to 250 kV/mm.

As shown in Fig. 2 (a), a small amount of positive homo space charge is observed. The shape of the space charge distribution is not packet-like but it distributes as it permeates into bulk of the sample. The permeated edge of the distribution is about 20 μm from the anode electrode. On the other hand, in the case of 250 kV/mm, such homo space charge is not observed while a small amount of hetero space charge is observed near cathode side. It is curious that the positive charge injection from anode is observed under low electric field while it is not observed under high applied electric field. To show the dependence clearly, figure 2 (c) shows a comparison of space charge distributions under various electric fields. Since the initial induced charge density on the electrode is proportional to the applied electric field, it is difficult to compare the shapes of the charge distributions. Therefore, the distributions Fig. 2 (c) are normalized using the peaks of the negative charge densities on cathode. Referring to Figure 2 (c), it can be seen that with the increase in the applied electric field, the edge of the distribution becomes closer to the anode surface. In other words, the charge injection suppression effect in LDPE/MgO nano-composite is more effective under higher applied electric field.

3. Induced dipole polarization

Authors have proposed that the filler added to base resin acts as an "induced dipole" under dc electric field and it works as a trap site in LDPE [12]. This phenomenon is derived form the difference of permittivity between base resin and filler. Figure 3 shows schematic models of electric field distributions around the filler [12]. In this figure, E_0 and σ_p stand for an applied electric field and induced charge density, respectively, on the surface of the filler. Below the figures of electric field distributions, the electric potential distributions around the filler are also described. In this figure, $V_0(r)$ and $V(r)$ stand for a potential distribution of applied voltage and distorted potential distribution by the existence of the filler ("potential well distribution"). In the case of figure 3 (a) when the permittivity of the base resin (ε_1) is same as that of filler, no charge is induced at the interface between them. In this case, the electrical potential distributions $V_0(r)+V(r)$ and potential well distributions $V(r)$ are flat. On the other hand, in this case of dielectric filler has higher permittivity (ε_{r2}) than that of base resin as shown in Fig. 3 (b), dipole charge σ_p is induced on the interface of the filler according to the following equation;

$$\sigma_p(\phi,\theta) = \varepsilon_0 E_0 \left(\frac{\varepsilon_{r2} - \varepsilon_{r1}}{2\varepsilon_{r1} + \varepsilon_{r2}} \right) \cos\phi \sin\theta \qquad (1),$$

where ϕ and θ stand for the angles in spherical-coordinate system. This equation is written as the filler has a complete spherical shape. In the model shown in Fig. 3, the direction of applied electric field E_0 is the direction with angles of $\phi = 0$ and $\theta = +90$ degree. As a result, negative charge induced left side, and same amount of positive charge induced other side. The potential well distributions $V(r)$ is calculated from difference between electrical potential distributions $V_0(r)+V(r)$ and applied electric potential $V_0(r)$. As shown in this figure, it is found that the electric field around the filler is distorted and a potential well is generates around the filler. Figure 3 (c) shows that the filler is conductive. In this case, the potential distribution has distorted much larger than that around the dielectric filler. We thought case (b) phenomena occurred in LDPE/MgO nano-composite. Since the potential well is generated around the dielectric filler under dc applied electric field, the injected carriers must be trapped around the filler.

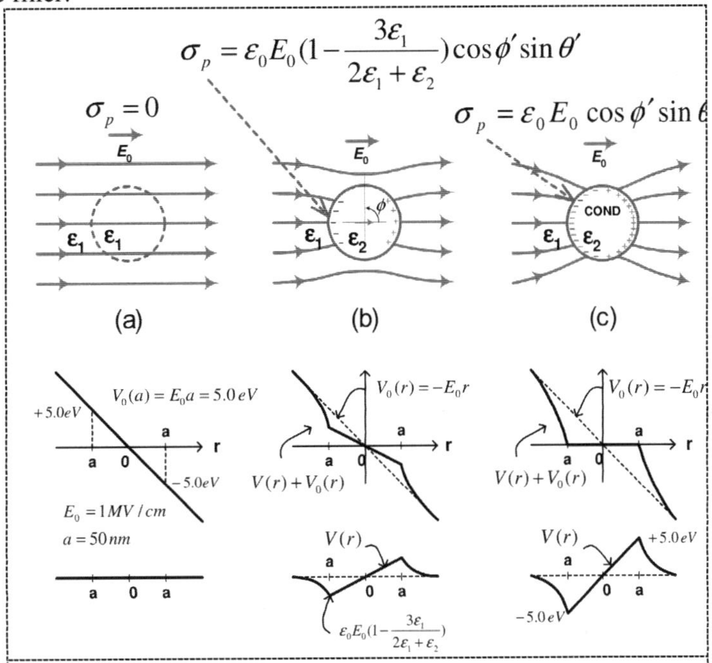

Figure 3. Upper figures: induced dipole electric charge and line of electric force distributions around a spherical dielectric filler with (a) the same permittivity as base resin, (b) different permittivity from the base resin ($\varepsilon_2 > \varepsilon_1$) and (c) conductor under applied electric field E_0. Lower figures: electrical potential distributions $V(r)+V_0(r)$ and potential well distributions, $V(r)$.

4. Electric field and potential distortion around nano filler

Here, a practical calculation of the potential distortion around MgO nano filler in LDPE is carried out. Figure 4 shows the electric field distribution around MgO filler and an electron. In this calculation, the electron is supposed as a charged defect or impurity included in LDPE. The distributions are produced by assuming that, for example, the MgO has a spherical shape with diameter of 50nm under applied electric field of dc 100 kV/mm. The values of 2.3 and 9.8 are used as the relative permittivities of

LDPE and MgO in this calculation. Since the effective cross-section area for the carriers passing around the MgO is much larger than electron's, the potential well formed around MgO is able to capture the carriers than that by the charged defect or impurity. Furthermore, the potential well around MgO collects many carriers while the electron is able to capture only one positive charge. The applied electric field depends on the effective cross-section area. In other words, when the higher electric field is applied, the larger area is effective for trapping the carriers around the filler.

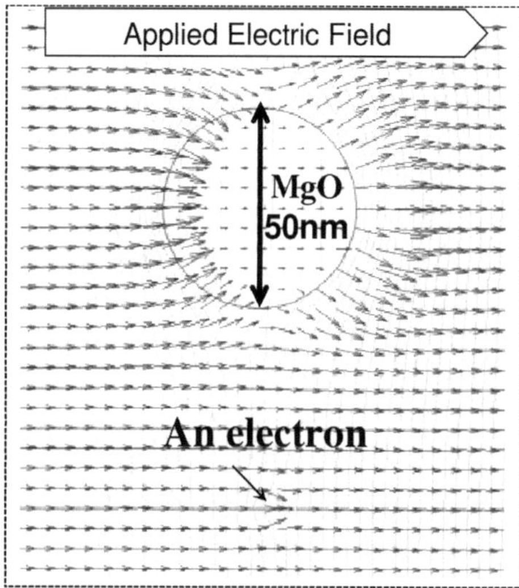

Figure 4. Electric field distortion around MgO filler and an electron produced by induced surface charge of a spherical MgO (diameter 50nm) under dc 100kV/mm and an electron.

To show the image of effective cross-section area for trapping clearly, a 3D schematic diagram of electric potential distribution is described in Figure 5. In this figure, two typical cases of three dimensional electric potential distributions around MgO filler are described. Figure 5 shows the potential distributions around MgO under (a) 100 kV/mm and (b) 250 kV/mm. In this figure, green and blue lines stand for the equipotential lines of negative 0.1 and 0.5 eV, respectively. It is easy to understand that the cross-section area of 0.1 eV formed around MgO under dc 250 kV/mm is much larger than that under 100 kV/mm. Furthermore, assuming MgO in the film shape material between parallel plane electrodes, the positive and negative potential wells are formed on both sides of MgO facing to the cathode and the anode sides, respectively. It means that the positive and negative carriers injected from anode and cathode are always captured with the negative and positive potential well formed around MgO filler, respectively. In other words, MgO filler under applied dc electric field acts as both positive and negative trap site.

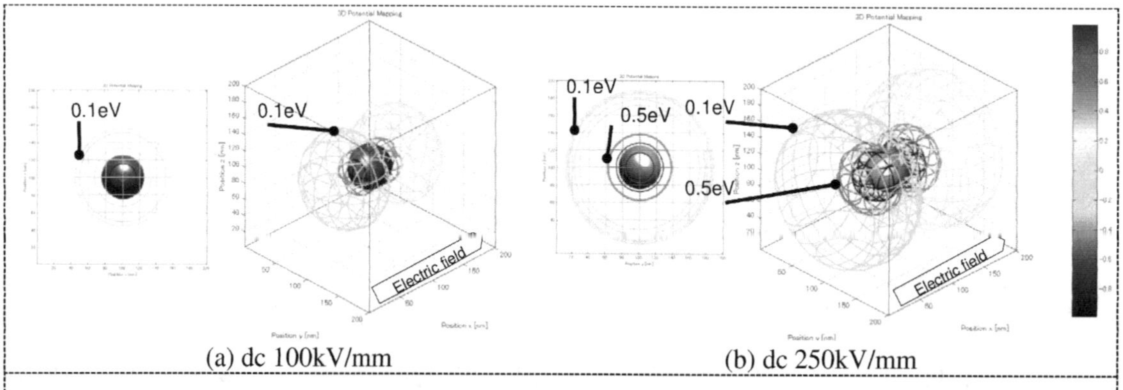

Figure 5. 3D electric potential well around MgO fillers produced by induced surface charge of a spherical MgO filler (diameter 50nm) under DC stress (a) 100kV/mm and (b) 250kV/mm.

5. Numerical analysis of suppression effects by nano fillers

Here, we try to simulate the charge distribution in LDPE/MgO nano-composite under various dc electric fields. In this simulation, we assume the cross-section area with potential of larger than 0.5 eV as the effective trapping area. For example, when the applied electric field is 100 kV/mm, the trapping area means the inside of the blue equipotential line shown in Fig. 5 (a). The calculated cross-section area of it is about 2700 nm^2. For example, when the content of MgO is 1 phr, numbers of MgO nano fillers is about 38 per 1 μm^3 and average distance between MgO fillers is about 333nm. In this simulation, we assume that the injected carrier move according to the electric field and it doesn't pass through the effective trapping area.

Figure 6 shows simulation results of trapping areas in LDPE/MgO (1phr). In this figure, each square shows a view of a layer in bulk of the sample with a distance of 1, 3 and 5 μm from charge injection surface. In each square, the white and black areas show where the carrier is captured by the well and where carrier can pass through the layer, respectively. It means that each white square pattern show the effective trapping area around MgO filler. Please note that the effective trapping area is shown using a square pattern for ease of calculation. Since the actual MgO filler is dispersed randomly, the white square pattern also put in each layer randomly. Here, since we assume the carrier can not pass through the effective trapping (white) area, the passing area (black) gradually decrease with increase the distance from the injection surface as shown in Fig. 6. It is found that the each size of the white square pattern under dc electric field of 50 kV/mm is smaller than that under 250 kV/mm. It means that the cross-section of the effective trapping area under 50 kV/mm is smaller than that under 250 kV/mm. Since the each size of the white square pattern under electric field of 50 kV/mm is small, there still remains black area even at the distance of 5 μm away form the injection surface. Under the electric field of 250 kV/mm, however, there is almost no such black area at the distance of 5 μm away from the injection surface. By the simulation, we can calculate the injection depth of the space charge from the injection surface. Figure 7 shows the calculated injection depth under various applied electric field. It is found that the injected carrier under electric field of 50 kV/mm can pass through the bulk of the sample for more than 20 μm while the injected carrier under 250 kV/mm cannot pass through even for 5 μm. From this simulation results, it is clearly found that the injected carrier can be mobile under lower applied electric field for longer distance from the injected surface. Furthermore, the simulation result seems to be close to the experimental result show in Fig. 2. The edge of the injected charge distribution under electric field of 50 kV/mm is located at the distance of about 20μm from the injected electrode and the distance becomes shorter with increase of the applied electric field. Since

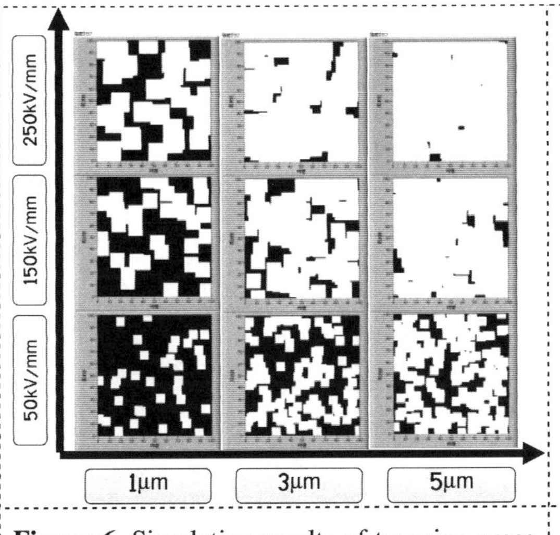

Figure 6. Simulation results of trapping areas in LDPE/MgO 1phr. Vertical axis: depth from electrode.

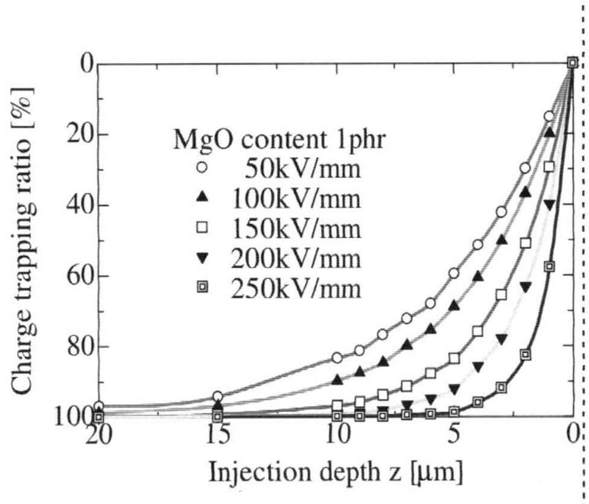

Figure 7. Space Charge trapping ratio by potential well. Vertical axis: injection depth from electrode.

the simulation was carried out based on parameters with rough assumptions, it is not a precise results. However, since we can have such very close simulation result to the experimental result even with such rough assumption, we think the basic proposed idea for the charge injection suppression effect by adding MgO nano filler to LDPE is a meaningful explanation.

6. Conclusion

Space charge formation in LDPE/MgO nanocomposite film subjected to a high electric field greater than 100kV/mm has been studied using PEA system and numerical simulation. When a high electric field was applied to LDPE, a huge amount of positive charge is injected into bulk and the charge accumulation distorts local electric field strongly. On the other hand, in LDPE/MgO nanocomposite film, which is including only less than a few wt% of MgO filler, no space charge formation is observed even under such severe applied electric field. To explain the difference in space charge formation, we proposed a trapping model of "induced dipole polarization". On the basis of this model, we showed the followings.

(1) Induced dipole polarizations of nanoparticles subjected to a high electric field are enough to become trapping site for carriers.
(2) Injection depth and the effective trapping by the induced dipole around the added nano filler depend on the difference between relative permittivities of nanoparticle and base resin. They also depend on applied electric field, too.
(3) Under more than dc electric field of more than 100 kV/mm, the lower applied electric field can make a longer space charge injection depth in LDPE/MgO (1 phr) nano-composite.
(4) The simulation for injection depth of injected charge shows a good agreement with the experimentally measured charge distribution using PEA system.

7. References

[1] Fuse N, Ohki Y, Kozako M and Tanaka T 2008 *IEEE Trans. DEI* **15** 161-169
[2] Roy M, Nelson K, MacCrone R K, Schadler L S, Reed C W, Keefe R and Zenger W 2005 *IEEE Trans. DEI* **12** 629-643
[3] Imai T, Sawa F, Nakano T, Ozaki T, Shimizu T, Kozako M and Tanaka T 2006 *IEEE Trans. DEI.* **13** 319-326
[4] Murata Y, Sekiguchi Y, Inoue Y and Kanaoka M 2005 *Proc. 2005 International Symposium on Electrical Insulating Vol.3* 650-653
[5] Murakami Y, Nemoto M, Okuzumi S, Masuda S, Nagao M, Hozumi N, Sekiguch Y, Murata Y 2008 *IEEE Trans. DEI.* **15** 33-39
[6] Li Y, Yasuda M and Takada T 1994 *IEEE Trans. DEI.* **1** 188-198
[7] Hozumi N, Takeda T, Suzuki H and Okamoto T 1998 *IEEE Trans. DEI.* **5**, 82-90, 1998.
[8] Matsui K, Tanaka Y, Takada T, Fukao T, Fukunaga K, Maeno T and Alison J M 2005 *IEEE Trans. DEI.* **12** 406-415
[9] Matsui K, Tanaka Y, Takada T and Maeno T 2008 *IEEE Trans. DEI.* **15** 841-850
[10] Hayase Y, Aoyama H, Matsui K, Takada T, Tanaka Y and Murata Y 2006 *IEEJ Trans. FM*, **126** 1084-1089
[11] Tanaka T, Kozako M, Fuse N and Ohki Y 2005 *IEEE Trans. DEI.* **12** 669-681
[12] Takada T, Hayase Y, Tanaka Y and Okamoto T 2008 *IEEE Trans. DEI.* **15** 152 160

Acknowledge
We are also grateful to Mr. Tatsuki Okamoto (CRIEPI) for constructive comments and discussions.

Breakdown and space charge formation in polyimide film under DC high stress at various temperatures

Y Kishi[1], T Hashimoto[1], H Miyake[1], Y Tanaka[1] and T Takada[1]

[1]Tokyo City University, 1-28-1, Tamatsutsumi, Setagaya-ku, Tokyo, JAPAN

E-mail: ytanaka@tcu.ac.jp

Abstract. Relationship between breakdown strength and space charge formation in polyimide film under dc high stress at various temperatures is investigated using pulsed electro-acoustic (PEA) method. Some typical results of the space charge observations show that hetero space charges are always found before breakdown. The amount of the hetero charges increase with increase of temperature or increase of applied electric field. Since the enhancement of the internal electric field in the sample by the accumulation of the hetero charges is not so large, the accumulation doesn't seem to be an immediate cause of breakdown. However since it is always observed before breakdown, it may be predictive information for breakdown. In a certain case, the breakdown occurs after voltage application for few hours. However, while we give an interval of short circuit condition after observing the hetero space charge under dc stress, the total voltage application time to breakdown is almost the same to the case without the interval. It means that the hetero space charge generation may show a kind of degradation of the material.

1. Introduction

Since a polyimide film shows a high dielectric strength even at high temperature, it is used as an insulating material for various electric and electronic devices, such as a flexible printed-circuit board in mobile phone, sheath of winding wire for motors, etc. Furthermore, recent current of downsizing and cost reduction demands it to be reduced its thickness. It means that the polyimide film is demanded to show a good performance as an insulating material under high electric field even under various sever condition, such as at high temperature with high humidity. Especially, a breakdown characteristic is one of important factors to estimate the material under such condition. On the other hand, the space charge accumulation sometimes strongly affects to dc breakdown strength. For example, so called, "packet-like charge" generation in low density polyethylene under dc high electric field of more than 100 kV/mm affects its breakdown strength [1]. However, the relationship between the breakdown properties and space charge accumulation under high dc electric field in other materials has not been clear yet. Therefore, we have been investigating the relationship in polyimide film. To investigate the relationship, the pulsed electro-acoustic (PEA) method [2] is used. The PEA method is useful technique to investigate the characteristics under high electric field [3] at high temperature [4, 5]. In this paper, authors introduce the space charge accumulation characteristics of polyimide film under dc high electric field near breakdown strength at various temperatures.

[1] Y Tanaka, Tokyo City University, 1-28-1, Tamatsutsumi, Setagaya-ku, Tokyo, JAPAN

2. Samples and experimental procedure

2.1. Samples
A commercially available polyimide film, "Kapton® type H" supplied from DU PONT-TORAY CO., Ltd, is used for all experiments. Nominally 125 and 75 μm thick samples are used for space charge measurement and breakdown test, respectively. The breakdown strengths at room temperature are described on a catalogue of merchandise as 207 and 339 kV/mm for 125 and 75 μm thick samples, respectively.

2.2. Measurement system
Figure 1 shows a PEA measurement system for high temperature [4, 5]. Since the system has a piezo-electric sensor of $LiNbO^3$, which shows a stable performance even at relatively high temperature, the system is applicable to the measurement condition up to about 100 °C. In an upper electrode unit, the sample is immersed into silicone oil to keep temperature stable and to avoid a flash over discharge. Temperature in the upper electrode unit is heated using a computer controlled heater put around the unit. The temperature is measured using a thermo couple put into the silicone oil. The maximum applied dc stress to the sample is about positive or negative 20 kV. Since the thicknesses of samples are nominally 125 and 75 μm, the maximum applied electric fields are about 160 and 260 kV/mm, respectively. The dc voltage is applied to the sample using computer controlled high voltage amplifier (HEOPT-20B10, Matsusada Precision Inc.). The dc high voltages are applied to the sample through a semi-con layer put on the sample. The semi-con layer is used to match the acoustic impedance of the sample to that of upper electrode [6]. The semi-con layer used for this experiment is made for a commercial XLPE power cable, and it is supplied by a cable company. Unfortunately, the detail information of the layer is undocumented. However, the surface of the layer is smooth enough to make a good contact with the sample and it is available even at high temperature of 80 °C. Since the sample is used without any evaporated metal electrodes on the surfaces, the interfacial condition at upper electrode is different from the lower one. The spatial resolution of the measurement using this system is about 10 - 20 μm depending on measurement temperature. It means that the relative spatial resolution for the sample with thickness of 125 μm is about 8 - 16 %. If we attempt to measure the space charge distribution in 75 μm thick sample, the resolution is not enough. That is reason why the result of space charge measurement for 75 μm thick sample is not reported on this paper. However, the breakdown test was also carried out using this system to match the measurement condition of it to that for the space charge measurement. The measurement principle and typical procedure are described elsewhere [2, 3, 6].

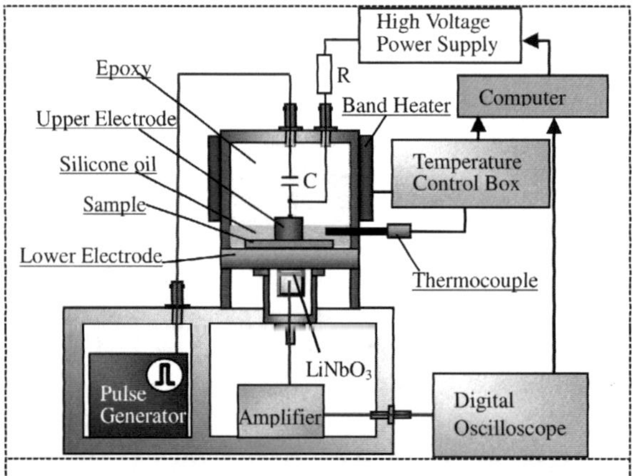

Figure 1. Measurement system

2.3. Measurement procedure
After putting the sample in the system, temperature is gradually increased with a heating rate of about 1 °C /minute up to a target temperature, and then the temperature is kept for at least 30 minutes. The target temperatures in the experiments are 30, 40, 50, 60 and 80 °C. After that, a target voltage is applied to the sample. Since a rapid voltage rise sometimes leads a flashover on the surface of the sample, the voltage is increased with a rate of 3 kV/s using a computer controlled power supply to avoid such flashover. When the voltage reaches to a target voltage, the voltage is kept for at least 2 hours if the breakdown doesn't occur within 2 hours. When the breakdown occurs, the voltage is immediately shorted automatically, and the measurement is finished with recording the time to breakdown. The target voltages depend on target electric fields and the sample thickness. In the experiments, the target electric fields are 110, 130 and 150 kV/mm for 125 μm thick sample, and 170, 190 and 210 kV/mm for 75 μm thick sample. The PEA measurements are carried out with every 5 seconds automatically using computer. The averaged signal made from 120 times shot of the pulse voltage is obtained as each datum. An adequate data processing including calibration process makes the voltage signal to a charge distribution. In addition to the above mentioned experiments, an interruption of voltage application and a repeated voltage application tests are carried out. Details of the tests will be mentioned later with the each result.

3. Results and discussion

3.1. Typical result
At first, a typical measurement result and how to display the results are introduced. Figure 2 (a) shows a typical measurement result of space charge distribution. This result is obtained by measuring 125 μm thick Kapton® film under applied electric field of 150 kV/mm at 80 °C. In this figure, firstly (just after beginning of voltage application) and finally (just before breakdown) observed data are drawn in red and blue lines. The left and right peaks show the induced charges on cathode and anode, respectively. As shown in figure 2 (a), positive charge is observed near the cathode electrode. Such kind of space charge accumulated near the electrode of opposite polarity is generally called a "hetero" space charge.

Figure 2 (b) shows electric field distributions in the sample at various times. The distributions are obtained by integral calculation using the measured space charge distributions shown in figure 2 (a). In this figure, the initial applied electric field has a flat shape and the value of the flat electric field is coincident with the applied electric field of 150 kV/mm. When the space charges accumulate in the bulk, the electric field is distorted by the charge accumulation. Therefore, the distribution has a maximum peak. In the case of measurement in LDPE, the maximum electric field show a significant meaning for its breakdown. Therefore, the maximum electric field in Kapton® is also calculated for each datum. As shown in figure 2 (b), the maximum electric field at just before breakdown (blue line) is about 190 kV/mm, and it means that the electric field is enhanced for about 27 % by the charge accumulation.

Using the PEA measurement, above mentioned profiles are originally obtained. However, since many time dependent profiles are drawn together in this figure, it is hard to recognise the time dependent properties. Therefore, authors use a colour chart, as shown in figure 2 (c), to show the measurement results clearly. The charge density is expressed using a colour bar beside the figure. In this figure, red and blue colours indicate positive and negative charge densities, respectively. The horizontal and vertical axes stand for position and time after voltage application. In this case, breakdown was observed at 11 minutes after beginning of voltage application. It is found that small amounts of positive and negative charges appear near anode and cathode electrodes, respectively, immediately after the voltage is applied to the sample. Such charges accumulated near the same polarity to the electrodes are "homo" space charges, and the charges are recognised, in many cases, as the injected charges from electrodes. On the other hand, positive and negative "hetero" space charges are also observed about few minutes after the beginning of voltage application. The amount of hetero charges gradually increases, and then breakdown occurs in the sample. Since the colour chart is convenient to show the time dependence as shown in figure 2 (c), the space charge measurement results are shown using the colour charts in following explanations.

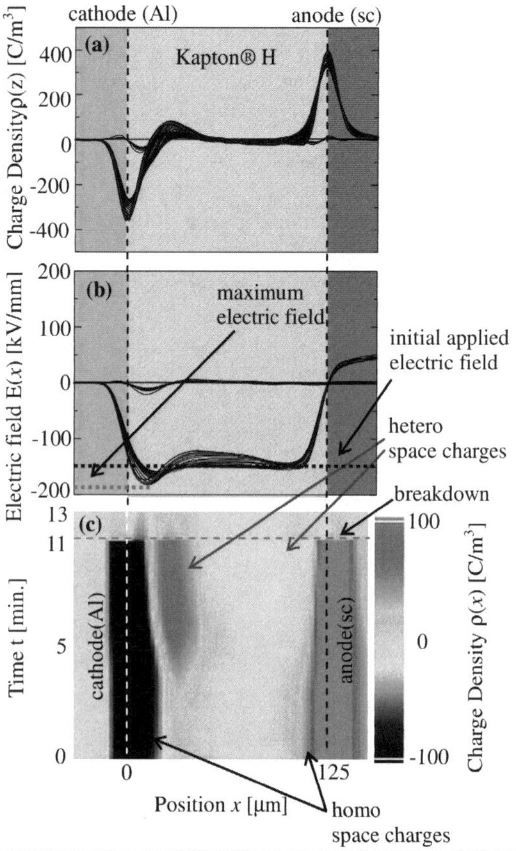

Figure 2. Typical measurement results of (a) space charge profile, (b) electric field distribution and (c) time dependent space charge behaviour in Kapton® under applied electric field of 150 kV/mm at 80 °C.

3.2. Electric field and temperature dependences of the space charge behaviours

Figure 3 shows applied electric field and temperature dependences of the space charge behaviours in Kapton®. The results are displayed in orders of applied electric fields and temperatures in vertical and horizontal axes, respectively. The electric fields of 110, 130, 150 and negative 110 kV/mm are applied to the 125 μm thick sample at 30, 40, 50, 60 and 80 °C. Measurements were carried out for at least 2 times in each condition and one of them is displayed here. In the cases that any remarkable space charge and a breakdown were not observed at least within 2 hours (240 min.), the results are shown with a sign of "without BD". It is clearly found that the breakdown is not observed under the electric field of 100 kV/mm at 30, 40, 50 °C, breakdown is not observed at least within 2 hours. On the other hand, under the field of 110 kV/mm at 60 °C, the breakdown is observed at 200 minutes later. In this case, positive hetero charges are observed from about 100 minutes later and its amount of charges gradually increase, and then finally the breakdown occurs. A small amount of hetero negative charge also observed before the breakdown. At 80 °C, such hetero space charges are also observed and the breakdown occurs at 120 minutes later. It is clear that, at the higher temperature, the larger amounts of hetero charges appears earlier before breakdown, and the breakdown also occurs earlier. When the electric field of 130 kV/mm is applied to the sample, the breakdowns occur at 213, 133, 83 and 50 minutes later at 40, 50, 60 and 80 °C, respectively, while it is not observed for 416 minutes at 30 °C. In the case that the breakdown occurs, the hetero space charges are observed before the breakdown. Under the electric field of 150 kV/mm, the breakdown is observed at all temperatures at earlier time with larger amounts of hetero space charges before breakdown. Judging from these results, under the higher electric field at the higher temperature, the breakdown occurs at the earlier time after the beginning of the voltage application with having the larger hetero space charges before the breakdown. When even the negative electric field of 110 kV/mm is applied to the sample, the characteristics of the breakdown and the space charge accumulation are almost the same as the case of positive electric field application. However, the amount of the negative hetero space charge is smaller than that observed under positive electric field application. Anyway, an especially remarkable property in these results is the hetero space charges are always observed before the breakdown.

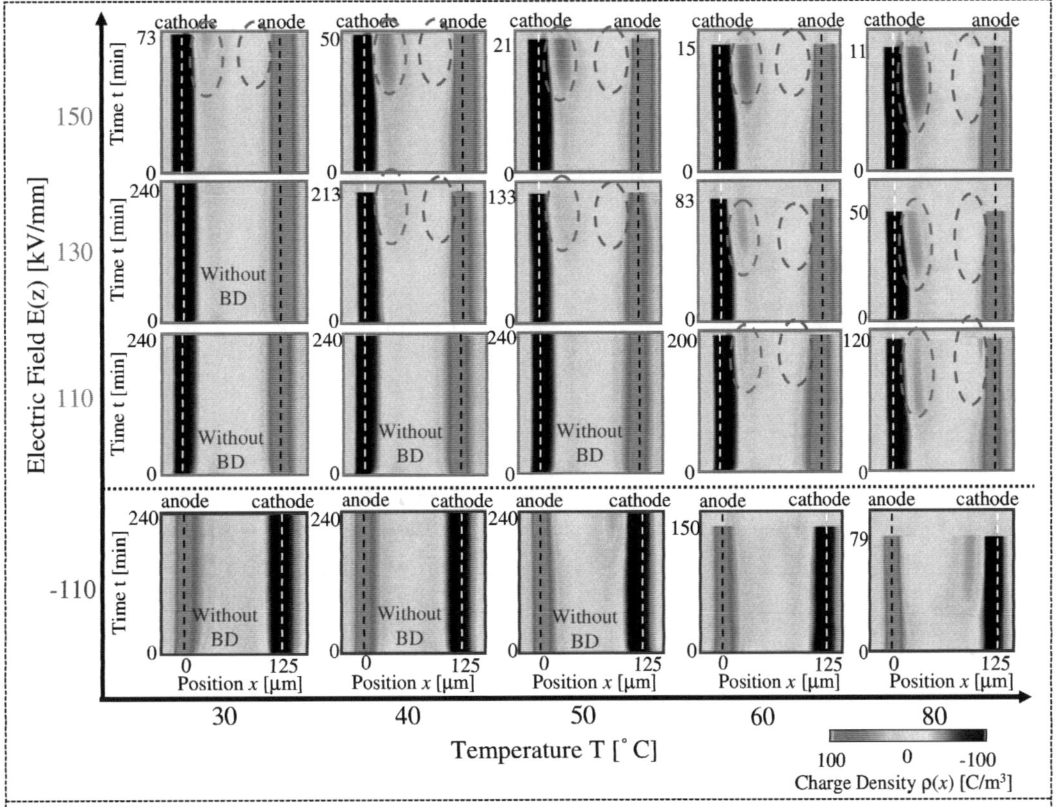

Figure 3. Space charge behaviours in Kapton® under various electric fields at various temperatures.

3.3. Relationship between temperature and time to breakdown

Figure 4 shows a relationship between temperature and time to breakdown. In this figure, the data obtained using 75 μm thick samples are also plotted. All data are averaged value form at least two times measurements. It is found that the time to breakdown shows a linear decrease with increase of temperature under the applied electric field. Since the vertical axis is described using logarithm scale, the time to breakdown decreases exponentially with increase of temperature. Most of all approximated fitting lines show the almost same slop although there is exception. Unfortunately, the data taken under electric field of 110 kV/mm is too short to estimate the above mentioned point. It is necessary to further detailed investigation. However, it is clear when the sample is used at the higher temperature under the higher dc electric field, the risk of breakdown increases within the shorter time.

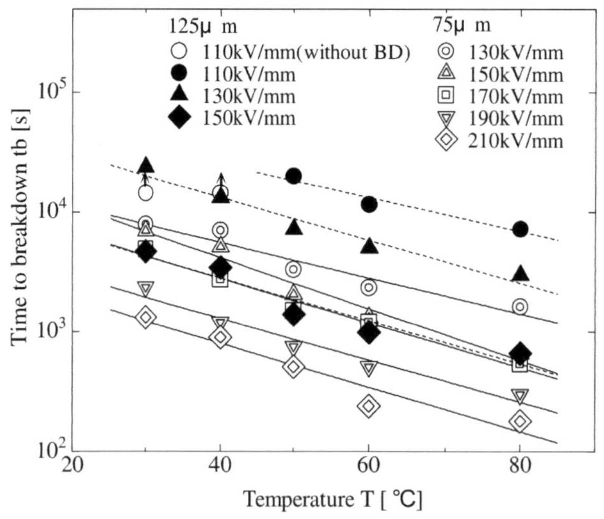

Figure 5. Relationship between temperature and time to breakdown.

3.4. Relationship maximum electric field and temperature

Figure 6 shows the relationship between the maximum electric field just before the breakdown and the temperature. Judging from the results, there seems be no significant relationship between them. In the case of LDPE, the breakdown seems to occur at almost the same as the enhanced maximum electric field, and it shows a direct affection of the space charge to the breakdown phenomena. On the other hand, there is no such property in this sample. It doesn't seem that the enhancement of the electric field is high enough to affect the breakdown strength. It means that the space charge formed in the Kapton® is not immediate cause of the breakdown but it may be a result of degradation for breakdown.

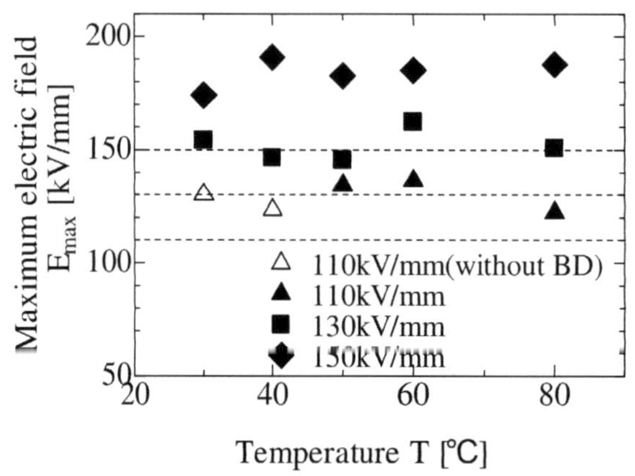

Figure 6. Maximum electric field just before the breakdown at various temperatures under various applied electric field.

3.5. Interrupting and repeating voltage application test

To show the space charge accumulation is a result of irreversible degradation, the interrupting voltage application test is carried out. In this experiment, the voltage is applied to the sample (130 kV/mm) at 60 °C for 50 minutes, when the hetero charge accumulation is observed clearly. Then the circuit is once shortened. After the interval of the short circuit condition for 12 hours, the same voltage is applied to the sample again. The result is shown in figure 7. As shown in figure 7, the accumulated hetero space charge finally disappeared at 12 hours later after circuit is shortened. However, the hetero space charge appears again immediately after the second application of the voltage while it takes for about more than 40 minutes to observe it clearly in the first voltage application. It means that the hetero space charge accumulation depends on the history of the high dc voltage application. After about 42 minutes in the second voltage application, the breakdown occurred. As shown in figure 3, it takes about 83 minutes for the breakdown to be observed with a continuous voltage application (130 kV/mm) at 60 °C. It seems that a total of the first and the second voltage application time to breakdown (50 + 42 = 92 minutes) is close to that of the continuous voltage application (83 minutes).

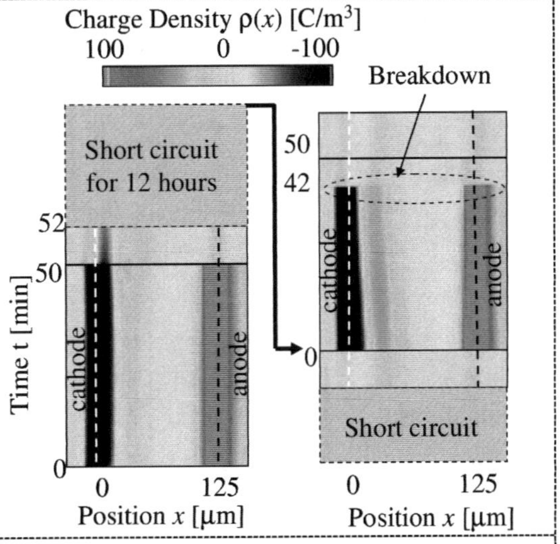

Figure 7. Space charge observation with interruption of voltage application.

To show the consequence of dc high voltage application history on the hetero space charge accumulation, a repeating voltage application test is also carried out. In this experiment, the voltage is applied to the sample repeatedly with interval of short circuit condition. At first, the voltage (130 kV/mm) is applied to the sample at 60 °C for 10 minutes, and then the circuit is shortened for 12 minutes. The same voltage is applied again after the short circuit interval for 10 minutes, and then the circuit is shorted again. Such a cycle is repeatedly carried out until the breakdown occurs. Figure 8 (a) shows the result of the repeated voltage application test. In this figure, only the observations under the voltage application are connected. For ease in comparison, the continuous measurement result under the same condition is shown in figure 8 (b). As shown in the figures, the short circuit interval procedure doesn't affect the space charge accumulation process at all. Judging from the result, the high voltage application even for 10 minutes may give a kind of irreversible damage to the sample and the damage accumulation may finally leads the breakdown. Furthermore, the appearance of the hetero space charges may be observed as a result of the damage accumulation. Since the experimental results are not enough for a statistic analysis of breakdown characteristics, it is hard to say the time to breakdown numerically. Further experimental works, however, may show the relationship between the amount of accumulated hetero space charges and time to breakdown, clearly. Another interesting point of view is how short term of the dc voltage application gives the damage to the sample. Polyimide film is sometimes used for an insulating material which is applied with short, several tens ms order, high voltage pulses intermittently. For example, in inverter driving devices, the insulating material is exposed to high voltage of inverter surge intermittently. If even such short pulse voltage gives the damage to the polyimide film, the damage gradually accumulates and finally it leads the breakdown. Therefore, we think that further investigation of the hetero space charge accumulation in polyimide film under high dc voltage is important to understand the characteristics of the actual devices.

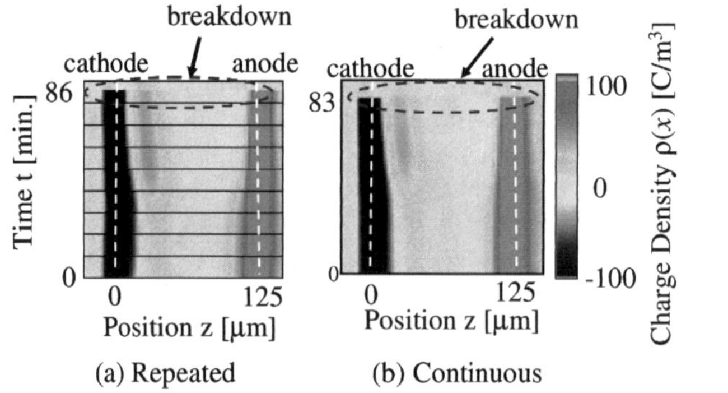

Fsigure 8. Space charge accumulation behaviour under dc high stress (a) with repeated intervals of the short circuit condition and (b) under continuous voltage application.

4. Conclusion

Space charge accumulation characteristics in polyimide film under high dc electric field at various temperatures are investigated using pulsed electro-acoustic (PEA) method. The results are summarized as follows:

(1) Hetero space charges are always observed before breakdown occurs in the sample.
(2) Under the higher applied electric fields at the higher temperature, the larger amount of hetero space charges are observed earlier after the voltage application.
(3) Voltage application with a long interval of short circuit condition doesn't recover the space charge accumulation properties. It shows the irreversible change of electric property must happen in the sample.
(4) Intermittently voltage application with relatively short duration (10 minutes) also leads the irreversible change of electric property in the sample.

5. References

[1] Matsui K, Tanaka Y, Takada T and Maeno T 2008 *IEEE Trans. DEI.* **15** 841-850
[2] Li Y, Yasuda M and Takada T 1994 *IEEE Trans. DEI.* **1** 188-198
[3] Matsui K, Tanaka Y, Takada T, Fukao T, Fukunaga K, Maeno T and Alison J M 2005 *IEEE Trans. DEI.* **12** 406-415
[4] Taima J, Inaoka K, Maezawa T, Tanaka Y, Takada T and Murata Y 2006 *Annu. Rep. CEIDP*, 302-305
[5] Maezawa T, Taima J, Hayase Y, Tanaka Y and Takada T 2007 *Annu. Rep. CEIDP*, 271-274
[6] Takada T, Tanaka T, Adachi N and Qin X 1998 *IEEE Trans. DEI.* **5** 944-951

Acknowledgments

A part of this research work was carried out in the 2nd Stage Knowledge Cluster Initiative No. 23 supported by the Ministry of Education, Culture, Sports, Science and Technology of Japan.

Effect of applied DC voltages and temperatures on space charge behaviour of multi-layer oil-paper insulation

Chao Tang[1, 2], G Chen[2], M Fu[3] and Rui-jin Liao[1]
1 The State Key Laboratory of Power Transmission Equipment & System Security and New Technology, Chongqing University, China
2 School of Electronics and Computer Science, University of Southampton, UK
3 AVERA T&D Technology Centre, Stafford, UK

E-mail: tangchao_1981@163.com

Abstract. In this paper, space charge in a multi-layer oil-paper insulation system was investigated using the pulsed electroacoustic (PEA) technique. A series of measurements had been carried following subjection of the insulation system to different applied voltages and different temperatures. Charge behaviours in the insulation system were analyzed and the influence of temperature on charge dynamics was discussed. The test results shows that homocharge injection takes place under all the test conditions, the applied DC voltage mainly affects the amount of space charge, while the temperature has greater influence on the distribution and mobility of space charge inside oil-paper samples.

1. Introduction

Insulating paper and oil are widely used as major insulation materials in high voltage direct current (HVDC) equipment, such as converter transformers, DC bushings and DC cables. The formation and dynamics of space charge under high electric fields will change the distribution of electric field inside the insulation material, greatly affecting the insulation performance. For example, trapped or low-mobility electrically charged species within the bulk can give rise to space charge, resulting in localized electric stress enhancement. This can cause further concentration of charge and lead to premature failure of the material [1, 2]. As the insulating paper inside the HCDV equipment takes the form of a multi-layer winding structure, a better understanding of charge dynamics becomes increasingly important in terms of the specific variation of the charge distribution in multi-layer insulation system.

Most of the discussion in the literature has concerned materials such as polymers and ceramics. Compared to measurements in polymers, the measurement of space charge in oil-impregnated materials is much more difficult. A few papers have reported the charge measurements in oil-impregnated insulating materials [3-6], such as space charge in impregnated pressboard and polypropylene laminated paper (PPLP). However, few have reported the space charge behaviour in multi-layer oil-paper insulation under different applied voltages and temperatures.

In this paper, using the PEA technique, a series of measurements were carried after subjecting the insulation system to different applied voltages and temperatures. Charge dynamics in the insulation system during the volts-on condition had been analyzed, and the influence of temperature on charge dynamics discussed.

2. Experiments

In this study, the insulation paper (Croylek Ltd.), has a thickness of ~ 60 μm for single layer. The insulation oil is Nytro 10X transformer oil, provided by Nynas Oil Company.

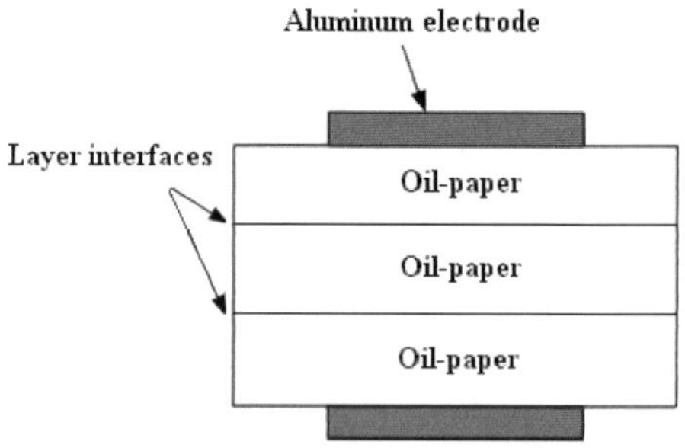

Figure 1. Schematic diagram of sample arrangement.

Before PEA tests, several pre-treatment steps were carried out upon samples. Firstly, the virgin samples were cut into round shapes with a diameter of ~ 2 cm. Then the paper samples were dried in a vacuum oven at 393 K (120 °C) for 20 min, reducing the moisture content to less than 0.2 %. After that, the paper samples were impregnated by fresh degassed insulation oil in sealed oven. As thin paper were used, the sample can be fully impregnated over a time of 24 h. As we know, oil-paper are usually used as multi-layer insulation in practical applications. Therefore, in this experiment, the PEA tests were performed on three layers oil-paper samples (~ 120 μm after oil immersed and being pressed by electrodes, in Figure 1). The samples were stressed at three different DC voltage levels (4 kV, 6 kV, 8 kV) and three temperatures (20 °C, 40 °C, 60 °C). Each time, an electric stressing time of 1 hour was tested.

3. Experimental results and discussion

3.1. PEA tests at 20 °C

Obviously from the space charge distributions shown in Figure 2, the cathode peak is sharp and evident; in contrast, the anode peak is wide and flat, this is because of attenuation and scattering of the acoustic waves by the oil-paper sample. As shown in Fig. 2 (a), (b) and (c), the peak value at the electrodes increases with the applied voltage from 4 kV to 8 kV. Under each applied voltage, because of homocharge injection at both electrodes, the amount of electric charge on both electrodes decreases as the stressing time increases. The positive charges accumulate in the vicinity of the anode, while the negative charges adjacent to the cathode. Charge was injected quickly after the voltage was switched on; an approximate equilibrium between positive and negative charge injection was observed after 30 min. No big difference noticed after that time. However, it may take a relatively long period (e.g. up to 10 000 s), to reach global equilibrium, i.e. the equilibrium between charge injection, extraction and transport [7].

The concentration of injected charge deepens into the bulk with the increase of the applied voltage. Under 4 kV, homocharge of each polarity penetrate no more than one layer. Under 6 kV, homocharge of each polarity enter more than one layer. It is possible that a small quantity of charge was neutralized in the middle layer. Under 8 kV, some homocharge reach the layer near to the opposite electrode.

Along with charge injection, the electrode peaks move slightly towards the centre of the sample, especially for the anode peak. There is less negative charge observed in the region adjacent to the cathode. There may be two reasons for this phenomenon. Firstly, compared to positive charge, the amount of negative charge injection did not increase as much as positive charge with the applied voltage. This indicates that negative charge may travel faster and be subjected to less blocking, reaching the anode quickly or being neutralized on the way. Secondly, the PEA technique only shows the net charge: the injected charge may be present together with the positive charge blocked at the interfaces.

Theoretically, in a PEA figure, the integral area of positive peaks is basically the same as that of negative peaks. However, because of the signal attenuation in oil-paper sample and the contribution of pulse, the test results are somewhat different from the theory. Besides, as a result of the deviation in paper thickness, the positions of electrodes shown in the test results are not exactly the same, which will not effect the final analysis. It is worth stressing that the shape of the charge profile image is not the same for the two electrodes, being broader on the electrode the further away from the detector (anode, right side), as a result of acoustic waves having travelled through the whole insulation thickness.

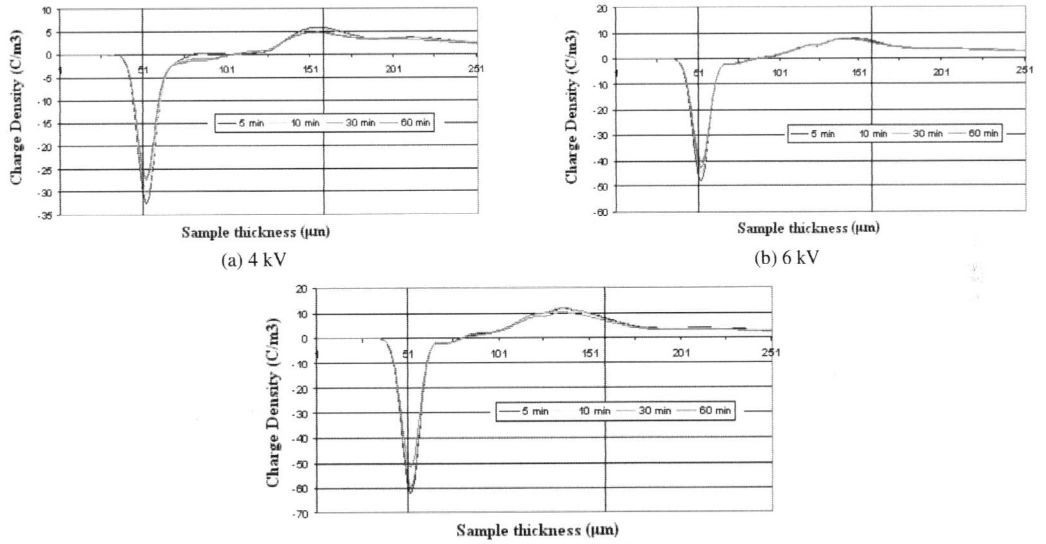

(a) 4 kV (b) 6 kV (c) 8 kV

Figure 2. Space charge dynamics at different DC electric fields (volts on at 20 °C)

3.2. PEA tests at 40 °C

When the test temperature was increased to 40 °C, the volts-on space charge dynamics at different DC electric fields are shown in Figure 3. As we can see, the temperature has a significant effect on space charge behaviour. Compared to 20 °C, the charge density on the electrodes increases at 40 °C. The homocharge injection takes place from both anode and cathode under three applied voltages.

At 4kV (Figure 3 (a)), homocharge injection from the electrodes takes place upon the voltage application, then charges move to the bulk of the oil-paper sample. After 5 minutes DC stressing, positive charges have already reached and began to accumulate at the oil-paper layer interface near to the cathode, and the peak value of charge density there is about 6.0 C/m^3 after 60 minutes DC stressing.

At 6 kV and 8 kV (Figure 3 (b), (c)), once the applied voltage was on, homocharge was rapidly injected into the oil-paper layer interface near to the cathode. More charges were injected compared to 4 kV. After 60 minutes DC stressing, the peak value of charge density in the oil-paper layer interface near to the cathode increased to 7.5 C/m^3 (6 kV) and 12.5 C/m^3 (8 kV).

The effect of temperature on charge dynamics is two-fold. On the one hand, the injected charges move quickly due to increased mobility. On the other hand, space charges gain more energy so that they can conquer the restraint of the first oil-paper layer interface near the anode and travel towards the second layer interface.

Apart from the charge magnitude, the distribution of space charge inside the oil-paper samples is hardly changed with the increase of the applied voltage.

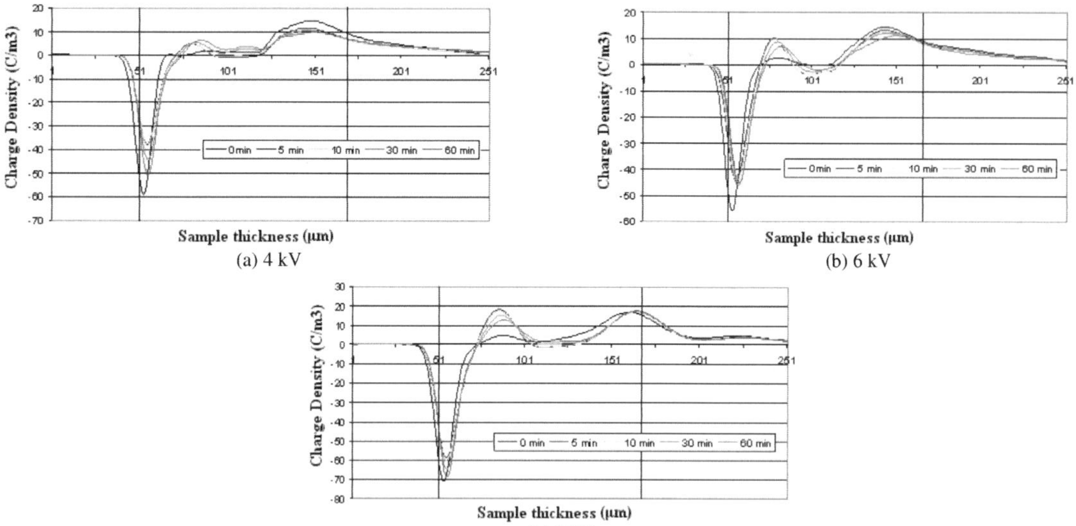

Figure 3. Space charge dynamics at different DC electric fields (volts on at 40 °C)

3.3. PEA tests at 60 °C

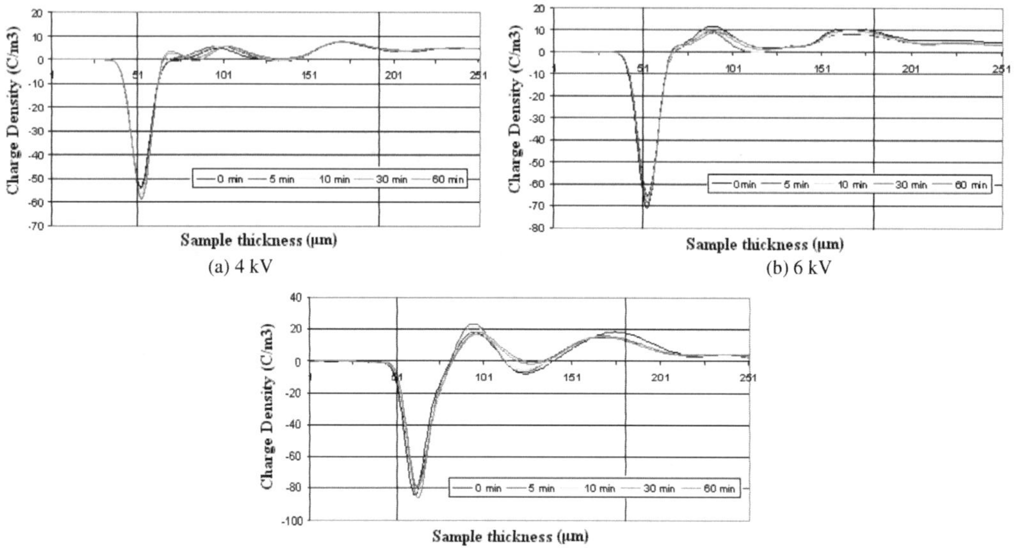

Figure 4. Space charge dynamics at different DC electric fields(volts on at 60 °C)

When the experimental temperature was increased to 60 °C, space charge measurements were done under different applied voltages (Figure 4). Similar to the situation at 20 °C and 40 °C, homocharge injection from both electrodes are observed. It is worth noting that the positive charges move extremely fast at 60 °C. Significant amount of positive charges have already accumulated at the layer interface near to the cathode once the voltage has been applied, and not much increase in the charge density was observed during the next 60 minutes of DC stressing. Measuring the time for the space charge to reach its saturation (steady-state equilibrium) distribution gives a practical method to describe the charge accumulation rate [8]. The increase in test temperature leads to the increase of the charge accumulation rate as it shortens the time for space charges to reach saturation.

4. Conclusions
Test results of PEA measurements on oil-impregnated paper samples were presented. Space charge dynamics under volts-on condition are analyzed. The following conclusions may be drawn:

(1) Once the external voltage is applied across the sample, homocharge injection takes place under all the test conditions. Positive charges are found to be dominant, and they tend to accumulate at the layer interfaces, increasing with applied voltage. This indicates that the oil-paper layer interfaces act as a barrier for positive charges. This will effect the distribution of electrical field and deteriorate the electrical behaviour of oil-paper insulation.

(2) The positive charges accumulate in different layers at different temperatures. At 20 °C, most of the positive charges accumulate in the layer near to the anode. When the temperature was increased to 40°C, charges move to the bulk and most of positive charges accumulate at the oil-paper layer interface near to the cathode. At 60 °C, both two layer interfaces have some positive charge accumulation.

(3) The amount of injected charge, charge mobility and charge distribution in oil-paper insulation are affected by the applied DC voltage and temperature. As far as the results showed in this paper are concerned, the applied DC voltage mainly effects the amount of space charge, while the temperature has greater influence on the distribution and mobility of space charge inside oil-paper samples.

5. References
[1] Zhang Y, Lewiner J, Alquie C and Hampton N 1996 Evidence of strong correlation between space charge buildup and breakdown in cable insulation *IEEE Transactions on Dielectrics and Electrical Insulation* **3**(6) 778–783

[2] Liu Z, Liu R, Wang H, Liu W 1989 Space charge and initiation of electrical trees *IEEE Transactions on Electrical Insulations* **24**(1) 83-89

[3] Rongsheng Liu and Wahlstrom G 1994 Measurements of the DC electric field in liquid impregnated pressboard using the pressure wave propagation technique, in *Conference Record of the 1994 IEEE International Symposium on Electrical Insulation* 130-136

[4] Nakagawa T 1999 Measurement of Space Charge Accumulation in PPLP in *Proceedings of the 1999 IEEE 13th International Conference on Dielectric Liquids* 533-536

[5] Mas P, Touchard G and Moreau O 1998 Space charge development measurements inside oil impregnated pressboard, in *1998 Annual Report on Electrical Insulation and Dielectric Phenomena Conference* 188-194

[6] Peter M and Marc J 1997 Space charge measurements on impregnated paper: a review of the PEA method and a discussion of results *IEEE Electrical Insulation Magazine* **13** 26-35

[7] Ciobanu R C, Pfeiffer W and Schreiner C 2003 Charge packet evolution in paper-oil insulation and derived technological considerations, in *Proceedings of the 7th International Conference on Properties and Applications of Dielectric Materials* **2** 695-698

[8] Mingli Fu, George Chen, Len A. Dissado and John C. Fothergill 2007 Influence of thermal treatment and residues on space charge accumulation in XLPE for DC power cable application *IEEE Transactions on Dielectrics and Electrical Insulation* **14**(1) 53-64

Acknowledgements
One of the authors (Chao Tang) wishing to acknowledge the Chinese Scholarship Council Funding for Joint Training PhD Student for the financial support provided.

The trapping characteristic of low density polyethylene in the presence of crosslinking by-products

Nuriziani Hussin and George Chen

School of Electronics and Computer Science
University Of Southampton, SO16 1BJ, United Kingdom

Email: nh07r@ecs.soton.ac.uk

Abstract. The by-products of dicumyl peroxide (DCP) from the crosslinking process such as acetophenone, cumyl alcohol and α-methylstyrene are said to be the sources of space charge formation in XLPE cable due to deep traps in the chemicals. However, by using space-charge-experimental approach, it appeared that these chemicals show a different trapping nature. This paper is intended to present this approach. Additive-free low density polyethylene (LDPE) was used as base material so that each chemical can be tested individually. Space charge measurement was done using the pulse electroacoustic (PEA) method. All results were compared to the clean LDPE to identify the contribution of the chemicals to the trapping characteristic. The data collected supported that although the chemicals introduce charge in the insulator, the charge decay is extremely fast especially in the presence of α-methylstyrene. It is believed that the chemicals modify the trapping characteristic of LDPE so that more shallow traps are formed in the insulator.

1. Introduction

Crosslinked Polyethylene (XLPE) has been applied in low and high voltage cables due to its good insulating properties. These properties are low dielectric loss, thermal and chemical resistance, resistance to deformation and stress-cracking as well as improved tensile strength and modulus. Nevertheless, the deposition of by-products in the PE structure formed during the crosslinking process is said to be a drawback of this insulating polymer. Space charge accumulation that is found in XLPE has always been associated with the by-products that act as deep traps in the insulator [1,2]. With the space charge existing in the insulator, premature dielectric breakdown might happen due to local field enhancement, ionization and electromechanical energy storage [3,4]. The real effects of the by-product chemicals on the trapping characteristic of insulator will be revealed in this paper.

2. Experiment

All samples used in this project were supplied by GoodFellow Cambridge Ltd. The samples with a thickness of 180 μm were cut into circular film. For the space charge measurement, sample diameter used was 36 mm. In this project, LDPE was chosen over XLPE to eliminate any possibility of having more than two by-products in the sample, allowing for easy identification of the role of each individual by-product in the dynamics of space charge.

2.1. Part I

The sample was soaked in α-methylstyrene and acetophenone for 2 h at room temperature and in cumyl alcohol at 80 °C since the melting temperature for cumyl alcohol is 23 °C. For space charge measurement, PEA technique was employed with semiconductor (Sc) and aluminium (Al) as anode and cathode, respectively. A pulse voltage of 600 V with time duration of 5 ns was applied to the sample to generate an acoustic signal wave. 8 kV DC voltage was applied to the samples at room temperature for one hour. In this research, measurements during short circuit were conducted to capture the slow moving charges in the sample. Every 10 min, the samples were short-circuited and the volts off measurements were conducted 5 s after voltage removal. As soon as the readings were recorded, the samples were recharged again until the next 10 min. This procedure was repeated until one hour of total charging time. Afterwards, the decay of the accumulated space charge in specimen was observed for another extended hour.

2.2. Part II

Another three samples were soaked into chemicals that consisted of two by-products. There are three different combinations of by-products, which has been simplified in the Table 1. Similar procedure as mentioned in *Part I* was conducted to these samples.

Table 1. The mixture of by-product chemicals

Sample	Chemicals
A	LDPE in acetophenone plus α-methylstyrene
B	LDPE in acetophenone plus cumyl alcohol
C	LDPE in α-methylstyrene plus cumyl alcohol

3. Results

3.1. Part I- Single by-product chemical

A clean LDPE sample is used as reference to distinguish the trapping characteristic that is caused by the by-products chemical and the polymer structure. The volt off result can be observed in figure 1.

In a clean LDPE sample, both positive and negative charges are injected into the sample and trapped in the vicinity of electrodes. After 20 min of charging, more positive charges are observed in the sample bulk. At the same time, more negative charges are trapped near the electrode and increase the peaks height. When the by-products are introduced into the LDPE, the charge density profiles are altered. In acetophenone soaked LDPE sample (figure 1(b)), the domination of negative charges in the sample can be observed. In the vicinity of the positive electrode, the amount of positive charge decreases by time, indicating the amount of negative charges overcoming the numbers of positive charges. As the aging time increases, more negative charges drifted towards the anode and at the same time, positive charge is suppressed. As a result, the amount of positive charge gets lesser by time.

In cumyl alcohol soaked LDPE sample, heterocharges are observed during the first 10 min of charging. As more negative and positive charges are injected from the cathode and anode respectively, the heterocharges can no longer be seen and are replaced by the homocharges. Over one hour of charging, both positive and negative peaks grow equally although one could roughly see that negative charge is accumulated in the centre of the sample.

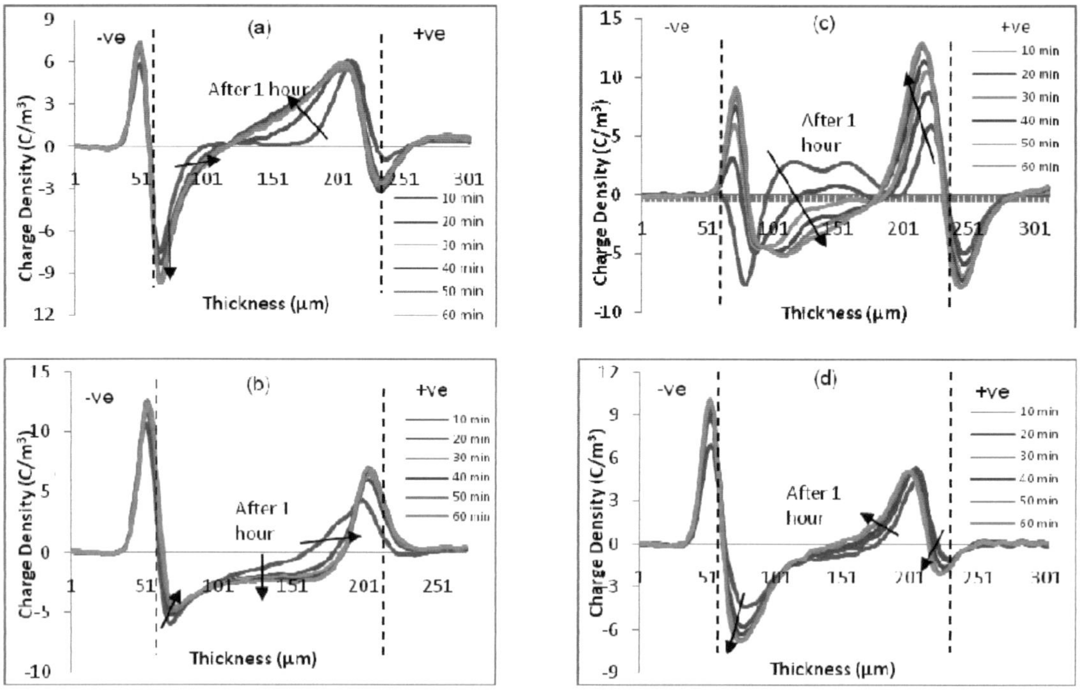

Figure 1. Charge Density of 180 μm a) clean LDPE b) acetophenone, c) cumyl alcohol and d) α-methylstyrene soaked LDPE stressed at 8 kV, during Volt OFF condition. Arrow shows the charge movement during one hour charging.

Unlike the former two by-products, α-methylstyrene has less effect on the charge density profile of LDPE in term of charge pattern. The space charge profile in figure 1(c) is almost similar to the charge profile in figure 1(a). However, lesser positive charge injection is seen in α-methylstyrene soaked LDPE thus equal amount of positive and negative charges are observed.

It is agreed that some of the results obtained in this paper are not similar to the results reported before. For instance, Maeno in his paper [5] found heterocharge in the presence of acetophenone and only homocharges are observed in cumyl alcohol soaked LDPE. These differences might be due to different electrodes system that was used during measurement. This factor is very vital in determining the charge injection profile in the dielectric sample. The electrode effects on space charge formation are reported elsewhere [6, 7].

3.2. Part II- Two by-products in sample

The charge density of Samples A, B and C for Volt Off measurement are presented in figure 2. For Sample A, the Volts Off measurement shows the domination of negative charges in the sample bulk similar to the charge density pattern of acetophenone soaked LDPE. Acetophenone is more prominent than α-methylstyrene in terms of controlling the charge build up. The positive charges are suppressed and as a result, a small positive peak is formed at the interface of the LDPE and anode. Similar characteristic is observed in figure 1(b).

The domination of acetophenone over the other by-product is also shown in Sample B. Although negative peak is only observed near to the cathode, the decreasing amount of positive charges by charging time and the movement of the positive peak towards the anode showed that the charges are suppressed. This observation is believed to be associated with the existence of acetophenone in the sample.

In the bulk of Sample C, no heterocharges appeared although cumyl alcohol subsisted in the sample. Positive charges are trapped at the vicinity of the anode and at the same time, the negative charges migrate from cathode to the sample bulk. As a result, more negative charge is seen in the sample. In Sample C, it is hard to tell which by-product is more dominant since the space charge dynamics does not show any distinctive pattern that can be associated to any of the individual by-products.

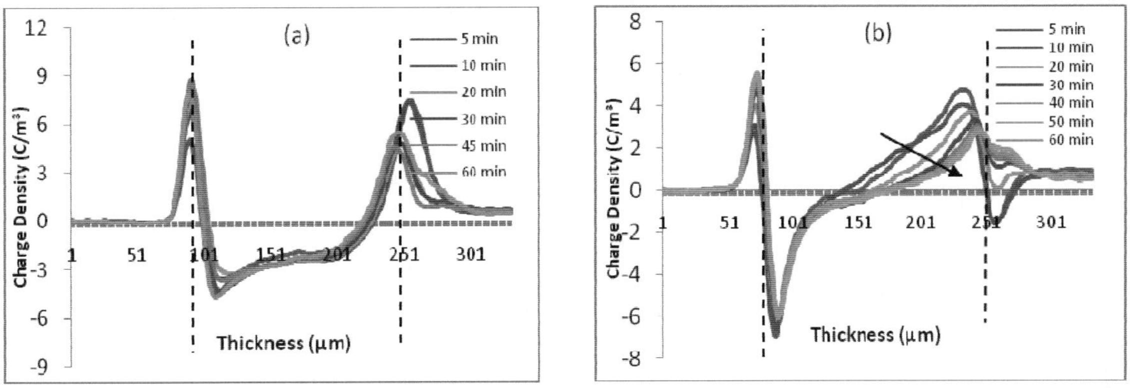

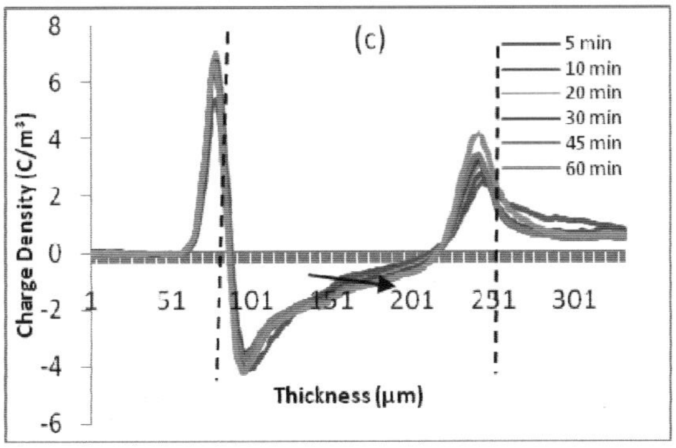

Figure 2. Charge density of LDPE soaked into (a) acetophenone + α-methylstyrene, (b) acetophenone + cumyl alcohol, (c) α-methylstyrene + cumyl alcohol, during Volts Off Condition. Arrow shows the charge movement during one hour charging.

4. Discussion

The total amount of charge in the sample, $q(x, t; Ea)$ that is relevant to the applied field E_a, is obtained by integrating the space charge density over the insulation thickness x, at a certain time t, [8].

$$q(t;E_a) = \frac{1}{L}\int_0^L |q_p(x,t;E_a)|\,dx \tag{1}$$

where 0 and L denote the positions of the electrode excluding the charges at the electrodes. The charge density profile of the samples in figures 1 and 2 could be analysed to obtain the total charge value during charging and decay process. The results of this calculation are presented in figure 3. Figure 3(a) demonstrates the total charge build up during charging process. The drop in the total charge that accumulated in cumyl alcohol soaked LDPE for 20 min of charging is due to the height reduction of heterocharges peaks by the injected charges. Nevertheless, as shown in the same figure, after 35 min of charging, cumyl alcohol soaked LDPE has more charges in the sample compared to the other samples. On the other hand, acetophenone and α-methylstyrene reduce the total charge injection and accumulation in the LDPE film. α-methylstyrene soaked LDPE has the least amount of charges in the sample.

Observation of the rate of decay in figure 3(b) can be used to investigate the trapping behaviour in insulator. Generally, the by-products gear up the decay process particularly in the first 15 to 20 min. Afterward, the charge decays gradually reduce to a rate that is similar to the rate of decay of the clean sample. Acetophenone and α-methylstyrene reduce the number of total charge trapped into deep trap. Although the total charge in cumyl alcohol soaked LDPE is greater than that of the clean LDPE, but the overall rate of decay for that sample is quite high.

Similar observation can be seen in Samples A, B and C. The total charge decay result indicated that any sample that contains α-methylstyrene will show a fast charge decay, compared to the one that does not contain the by-product, which is Sample B. Hence it can be said that α-methylstyrene is dominant in the charge decay mechanism although it does not give much effect in the space charge formation.

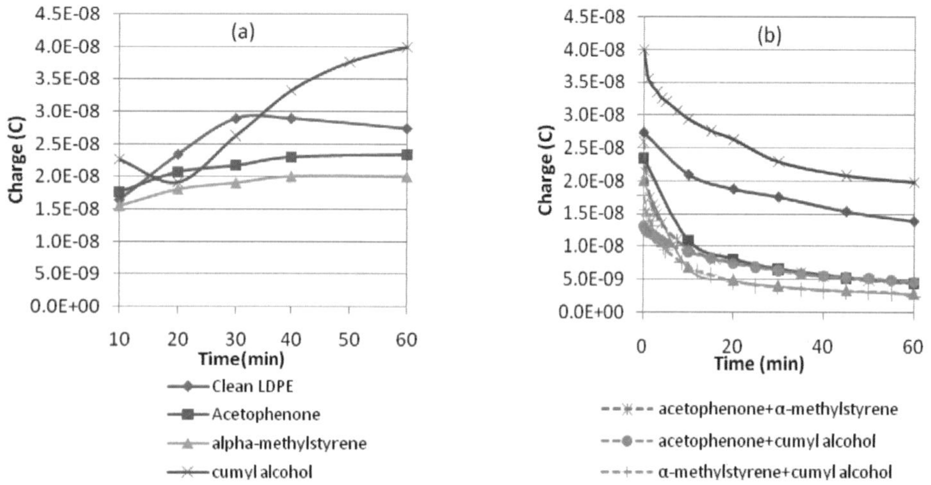

Figure 3. a) Total charge accumulated in sample during charging. b) Total charge decay in samples.

Based on the results, the by-products modify the trapping characteristic of LDPE in two possible ways which are, 1) by increasing the population of the shallow traps, and 2) by reducing the number of deep traps in the LDPE, replacing them with the shallow traps.

Voids and small spaces between the crystal structures that exist in the insulator will be filled by the byproducts. Originally, the regions of reduced density such as sub-microvoids with surrounding polymer chain will become deep electron traps [9]. Since the typical features of the chemical structures of these crosslinking by-products compared to LDPE are benzene ring, carbonyl group, double bond, and hydroxyl group, it is necessary to associate the changes in trapping behaviour with the existence of the chemical group [10].

The above argument comes with an assumption that during charging process, the charges have a higher tendency to fill in the shallow traps rather than the deep traps. The validity of this assumption has been confirmed via a short experiment involving LDPE sample that is charged at the same voltage, but for various charging periods. Samples that are charged for a shorter period of time will have faster charge decay. The result is presented in figure 4. Based on this assumption, it is proposed that the similar rate of decay observed in figure 3(b), during 20 to 60 minutes after short circuit in all samples is due to the original deep traps in LDPE.

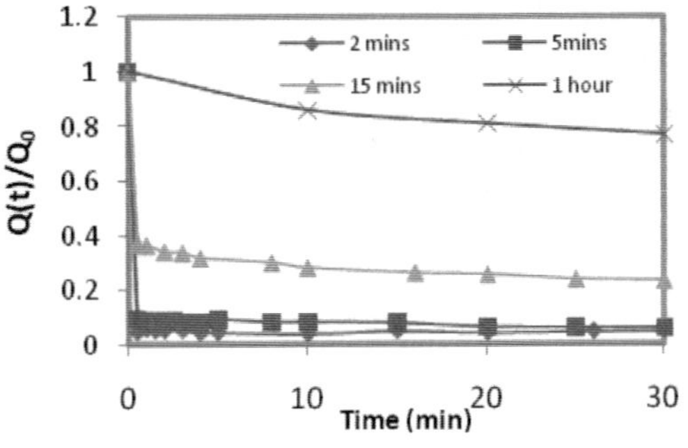

Figure 4. Plot of Q(t) in relative value with respect to Q_0=Q(0) against the decay time.

5. Conclusion

Our investigation upon the by-products showed that instead of introducing deep traps, the by-products increase the number of shallow traps in the insulator. This is due to the faster decay of the total charge in the presence to the crosslinking by-products.

α-methylstyrene has less effect on the charge density profile of LDPE but plays the main role in speeding up the decay process. Meanwhile cumyl alcohol is dominant in causing charge injection into the sample.

6. References

1. M. Meunier, N. Quirke, A. Aslanides. *Characterisation of Charge Carrier Traps in Polymeric Insulator* Conference on Electrical Insulation and Dielectric Phenomena. Volume Vol 1; 2000. p 21-24.
2. G. Teyssedre, C. Laurent, A. Aslanides, N. Quirke, L. A. Dissado, G. C. Montanari, A. Campus, L. Martinotto. *Deep trapping centers in crosslinked polyethylene investigated by molecular modeling and luminescence techniques.* Dielectrics and Electrical Insulation, IEEE Transactions on 2001;8(5):744-752.
3. Y. Zhang, J. Lewiner, C. Alquie, N. Hampton. *Evidence of strong correlation between space-charge buildup and breakdown in cable insulation.* IEEE Transactions on Dielectrics and Electrical Insulation 1996;3(6):778-783.
4. G. C. Montanari. *Electrical degradation threshold of polyethylene investigated by space charge and conduction current measurements.* IEEE Transactions on Dielectrics and Electrical Insulation 2000;7(3):309-315.
5. Y. Maeno, N. Hirai, Y. Ohki, T. Tanaka, M. Okashita, T. Maeno. *Effects of crosslinking by-products on space charge formation in crosslinked polyethylene.* IEEE Transactions on Dielectrics and Electrical Insulation 2005;12(1):90-97.
6. T. Hori, K. Kaneko, T. Mizutani, M. Ishioka. *Effects of electrode on space charge in low-density polyethylene.* Proceedings of the IEEE International Conference on Properties and Applications of Dielectric Materials; 2003; Nagoya, Japan. Institute of Electrical and Electronics Engineers Inc. p 855-858. (Proceedings of the IEEE International Conference on Properties and Applications of Dielectric Materials).
7. G. Chen, Y. Tanaka, T. Takada, L. Zhong. *Effect of Polyethylene Interface on Space Charge Formation.* IEEE Transactions on Dielectrics and Electrical Insulation 2004;11(1):113-121.
8. G. C. Montanari, D. Fabiani. *Evaluation of dc insulation performance based on space-charge measurements and accelerated life tests.* IEEE Transactions on Dielectrics and Electrical Insulation 2000;7(3):322-328.

9. J. P. Jones, J. P. Llewellyn, T. J. Lewis. *The Contribution of Field-Induced Morphological Change to the Electrical Aging and Breakdown of Polyethylene.* IEEE Transactions on Dielectrics and Electrical Insulation 2005;Vol 12 (No 5):951-966.
10. R. M. Naoshi Hirai, Toshikatsu Tanaka, Yoshimichi Ohki. *Chemical Group in Crolsslinking By-products Responsible for Charge Trapping in Polyethylene.* IEEE Transactions on Dielectrics and Electrical Insulation April 2003;Vol. 10 (Issue 2):320-330.

The effect of charge trapping on final thermally stimulated discharge currents in polyethylene naphthalate (PEN)

M Saidi[1] and N Saidi-Amroun[1]

[1]Material Physics Laboratory, Physics Faculty, University of Sciences and Technology (USTHB) BP 32 El-Alia, Bab-Ezzouar, Algiers 16111, Algeria

E-mail: amnadster@gmail.com (N Saidi-Amroun)

Abstract. The aim of the work presented in this paper was to analyse final thermally stimulated discharge current that appeared at the end of discharging process, when the temperature was increased from 100 to 180°C. This current exhibit two probable peaks. To fit the experimental data, two Gaussian components were used: the first one centred at 128°C near T_g and the second one at 150°C associated to pre-melting of crystallites. The sample (gold metallised PEN film 25μm thick) was submitted, prior to this process, to 72MV/m DC field during ten thousand seconds. The DSC analyse showed a T_g at 128°C and an endothermic peak at 150°C. These results imply that the trapped space charge cannot release from the sample and require an increasing of temperature. Above T_g, the molecular motion facilitates the de-trapping process.

1. Introduction

Due to their excellent electrical properties, polymers are widely used as electrical insulators. However, in particular conditions, especially under high electrical field, their insulating properties may be affected, as the space charge composed of trapped charge carriers, can have a drastic effect on dielectric properties, which influence tremendously the reliability of apparatus and systems where they are integrated. Indeed, this charge can increase the local electrical field and subsequently cause breakdown of the insulator.

The aim of this work was to make a correlation between the trapping of space charge and transient currents, by comparing the results obtained by three techniques. The material used in this study was the polyethylene naphthalate (PEN) which gives high performance films with excellent thermal mechanical and electrical characteristics.

Measurements were done in a large temperature-range including the glass transition temperature T_g so as to analyze the effect of polymer state, glassy or rubbery, on electrical properties.

2. Material

Polymers are currently used as dielectrics in a wide range of electro-technological applications. Since the polyethylene naphthalate (PEN) gives high performance films with physical, thermal and chemical characteristics better then those of polyethylene terephthalate (PET), it is used as a substitute for the latter in a large number of applications, particularly in film capacitors.

The PEN film of 25 μm thickness, and 1.36 g/cm^3 density was provided by Dupont de Nemours Luxembourg. The PEN sample was prepared with gold evaporated electrodes, 2 cm in diameter, on

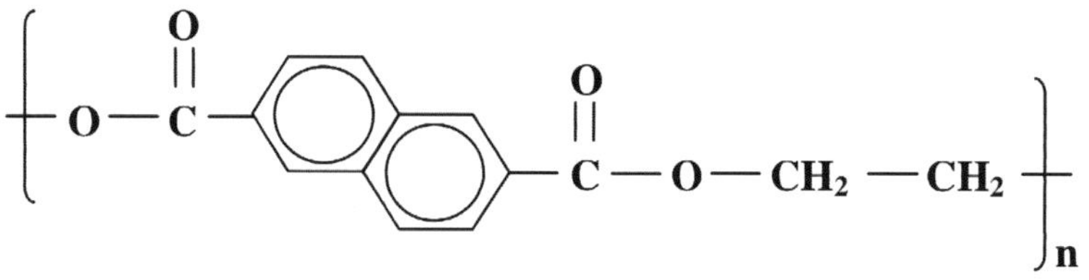

Figure 1. Chemical structure of PEN.

both sides of the sample. The chemical structure of PEN is reported in figure 1.

3. Experimental and method

3.1. Transient current technique
A DC voltage was applied to a sample at a constant temperature, and the charging current was measured up to three hours. The voltage was then removed and the discharging current of the short-circuited sample was recorded for three hours, according to the experimental setup [1]. The experimental cell was connected to a vacuum pump and a gas cylinder. This allowed to control the nature of the gas inside the cell, and to remove the moisture from the sample. A temperature controller connected to the heating element was located inside the cell. Hewlett Packard stabilized power supply (HP6525ADC) was used to deliver DC voltages up to 4000V. The current was measured by Keithley 617 programmable electrometer driven by a microcomputer. A high voltage relay switches from polarization when the sample was exposed to the voltage, to depolarization when the sample was grounded.

A new technique combining isothermal charging and discharging with non-isothermal discharging measurement proposed by Neagu et al [2], was used. In the second step of the experiment, the sample was heated at a constant rate of 2.4°C/min, from a charging temperature T_c to a final temperature T_f = 180°C. The current recorded was called "final thermally stimulated discharge current" or "non-isothermal discharging current".

3.2. DSC analyse
The thermal analysis of PEN samples was obtained by using the TA instruments (Differential Scanning Calorimeter, DSC Q10 t). DSC thermograms were recorded from 30 to 300°C by a constant heating rate of 10°C/mn under a nitrogen atmosphere to avoid thermal degradation. The obtained data was analyzed by TA Universal Analysis Software.

4. Isothermal currents

4.1. Charging Currents
Isothermal charging current, at different temperatures between 100 and 150°C, under a DC field of 72×10^6 V/m, are presented in figure 2. These results showed an anomalous behavior. Indeed, instead of monotonous decrease, a rise after few minutes of polarization [3] was observed, which was generally attributed to space charge effect.

Several works have reported the existence of these peaks with different polymers, like polyethylene (PE) [4-5], polyethylene terephthalate (PET) [6] and polypropylene (PP) [7]. In PEN very few [8] references cited this behavior. The charging current peak is similar to that given by Many and Rakavy [9] relating to transient SCLC (space charge limited current). It appears at time t_p :

$$t_p = 0.786.\frac{d}{\mu F} \qquad (1)$$

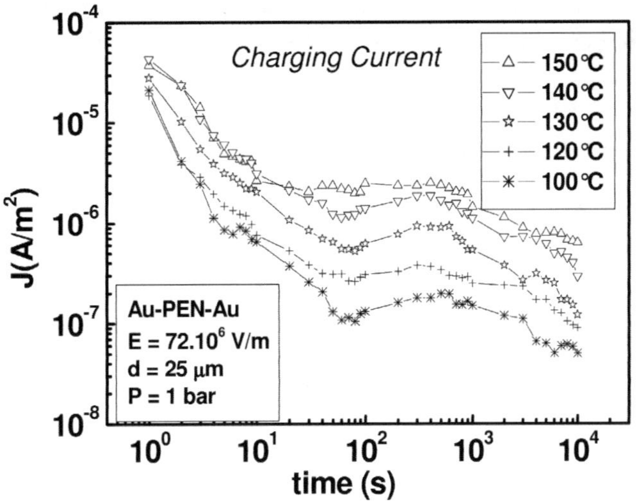

Figure 2. Isothermal charging current in gold metallised PEN at different temperatures.

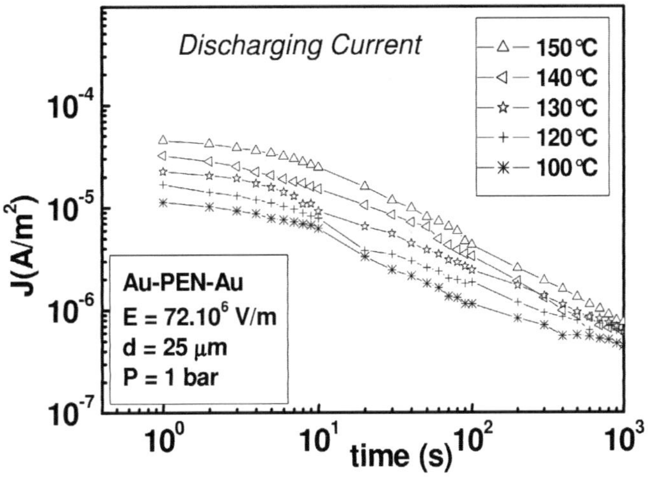

Figure 3. Isothermal discharging current in gold metallised PEN at different temperatures.

where F is the applied field, d the sample thickness, μ the carrier mobility and ε the permittivity of the polymer. This time was related to the transit time of a space charge between electrodes.

4.2. Discharging Currents

Figure 3 shows typical discharging current measurement subsequent to the charging current displayed in figure 2. It can be noticed that the discharging current present a monotonous decrease, suggesting a dependence on time given by the Curie-Von Schweidler law [9]:

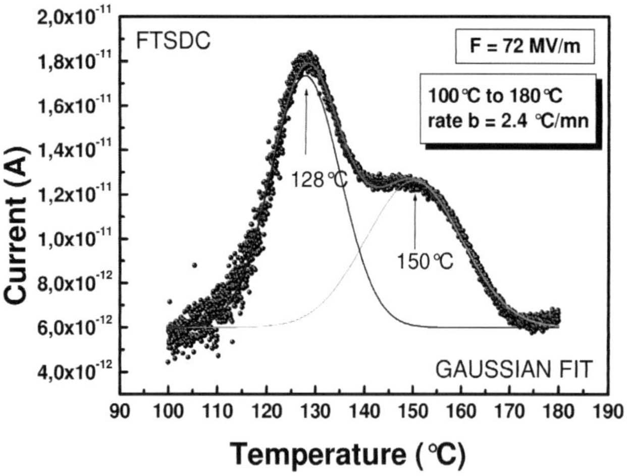

Figure 4. Non-isothermal discharging current in PEN between 100 and 180°C at a constant heating rate b = 2.4°C/min. Broken lines indicate the two Gaussian components used in the experimental data (+)

$$J = A\ t^{-n} \tag{2}$$

with n an exponent close to unity. Neagu [10] suggested that the isothermal discharging current described by this power law for the whole glass transition temperature range, is due to the space charge released from the deeper traps existing in the amorphous volume.
On the other hand, it should be noticed that the absorption current is not the mirror image of the discharging one. This suggests that the dipolar mechanism was not the predominant process. Mechanisms of dipolar origin are generally predominant at short time and low field [9].

5. Final thermally stimulated discharge currents

The isothermal discharging current (figure 4) becomes low but there is still space charge trapped in the sample. Subsequently the sample is heated up at a constant rate and the released current is recorded. Figure 4 presents the results obtained for a sample charged at 100°C under an electric field of 72 MV/m for 4000 s and isothermally discharged during 4000s. The non-isothermal discharging current was obtained for a heating rate of 2.4°C/min from 100 to 180°C. The variation of temperature as a function of time was linear. That allow us to represent the variation of current I as a function of temperature T. A high increase of the current was observed. An increase in temperature seems to release some of the trapped charges. Two peaks were obtained: the first one located at T_1 = *128°C* near T_g and the second one, which appears as a shoulder on the first one, at T_2 = *150°C*.

5.1. Analysis of I(T) spectra

The procedure used to fit the *I(T)* spectra is illustrated in figure 4. We have found that the experimental data can be properly fitted by a superposition of two Gaussian profiles: One described the release of space charge related with the relaxation in the glass transition temperature range, while the other takes into account the detrapped charge from deeper traps associated to the premelting of crystallites. The fitting parameters are summarized in table 1.

The charge involved in the discharging step can be calculated using equation 3:

$$Q = \int I.dt = \frac{1}{b}\int_{100}^{180} I.dT = \frac{A}{b} \tag{3}$$

Table 1. The Gaussian fitting parameters of a *I(T)* spectra (figure 4) corresponding to a 2.4°C/min heating rate.

Peak	Center T_i (°C)	Area *A* (A.°C)	Width (°C)	Height (A)	Charge *Q* (nC)
1	128	2.0 x 10^{-10}	14.3	1.1 x 10^{-11}	5.0
2	150	1.6 x 10^{-10}	19.7	6.6 x 10^{-12}	4.0

where *b* is the heating rate and *A* the Gaussian profile area.

Finally, the charge Q_1 was obtained by simple integration of the Gaussian profile peaked at $T_1=128°C$. With a similar procedure the charge Q_2, associated to the second curve, was also estimated. These charges are less then 1% of the charge $CV = 0.6\mu C$ on the capacitor for $V = 1800V$ and $\varepsilon_r = 3$.

5.2. *Heating rate effect*

The heating rate is generally an important parameter for non-isothermal measurements. It can give the possibility to investigate charge detrapping and transport mechanism.

With the aim to examine if the temperatures of the peaks are related the measuring conditions, the experiments were repeated in identical conditions but with higher heating rate.

The final thermally stimulated discharge current obtained with a sample charged at 100°C under an electric field of 72 MV/m for 4000s, and isothermally discharged during 4000s, for a heating rate of 4.8°C/min from 100 to 180°C, is displayed in figure 5. These experimental data can also be fitted by two Gaussian profiles, the first one peaked at 132°C and the second one at 151°C. The fitting parameters are reported in table 2

The first observation is that the heating rate does not affect significantly the appearance and the localization of peaks. On the other hand the height of the two peaks increases with *b*. Indeed, doubling the heating rate induced a more than doubling of the peak heights. But there is a less change in the charge values Q_1 and Q_2 as indicated in tables 1 and 2. In fact, the heating rate of 2.4°C/min is still high to permit the observation of the two peaks. Consequently, an increase of *b* will not affect largely the *I(T)* spectrum. Contrarily, the situation where *b* is increased from low values can exhibit a significant effect as mentioned by Neagu [11] in the case of PET films. It can also be noticed that the ratio of the height of the first peak to the second one is practically equal to ratio of the two heating rates, $\left.\frac{H_1}{H_2}\right|_{4.8°C/min} = 2.0$ and $\left.\frac{H_1}{H_2}\right|_{2.4°C/min} = 1.7$.

6. Differential Scanning calorimetry

The DSC thermogram of sample was given in figure 6. As the temperature increases, the glass transition, and the melting of PEN can be observed successively. The DSC thermogram of the semi-crystalline PEN shows a T_g (glass transition temperature) at about 128°C, and T_m (melting temperature) at 267°C. A small endothermic pre-melting peak was also observed at T_{pm} = 150°C. The presence of double melting peak was already reported [12] in PEN.

These experiments have allowed us to calculate the degree of crystallinity using the equation:

$$\chi(\%) = 100\frac{\Delta H_m - \Delta H_c}{\Delta H_f} = 44\% \qquad (4)$$

where ΔH_m *and* ΔH_c were respectively crystallization and melting enthalpy. $\Delta H_f = 103.4\ Jg^{-1}$ was a melting enthalpy of 100% crystalline PEN.

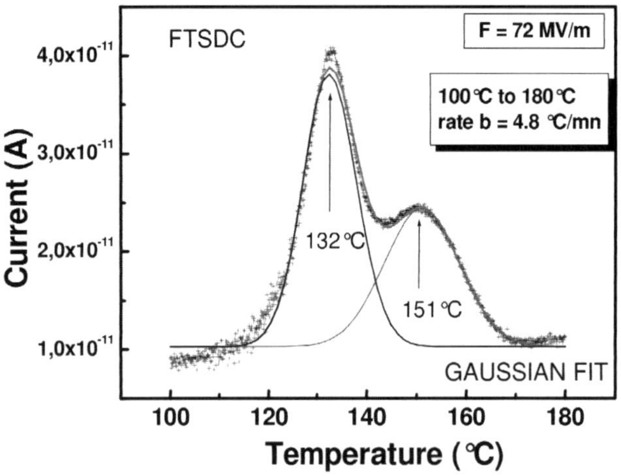

Figure 5. Non-isothermal discharging current in PEN between 100 and 180°C at a constant heating rate b = 4.8°C/min. Broken lines indicate the two Gaussian components used in the experimental data (+)

Table 2. The Gaussian fitting parameters of a $I(T)$ spectrum (figure 5) corresponding to a 4.8°C/min heating rate.

Peak	Center T_i (°C)	Area A (A.°C)	Width (°C)	Height (A)	Charge Q (nC)
1	132	3.8 x 10^{-10}	10.9	2.8 x 10^{-11}	4.8
2	151	2.6 x 10^{-10}	15.0	1.4 x 10^{-11}	3.3

7. Discussion

The increase of a non-isothermal discharge current is related to detrapping of space charge. The release of charges may be a result of increased motion of the polymer chains at the higher temperature, especially near T_g. Indeed, it is well known that glass transition temperature is associated with chain segment cooperative motion induced by diffusive rearrangement of main chain portions [13].

The peaks are localized at two temperatures which are in good concordance with DSC analysis and current-field $J(F)$ characteristics presented in previous works [14]. It would be noticed that this temperature (150°C) was presented by industrials as the maximal temperature of use. The two peaks suggest that there are, at least, two sorts of traps in which charges may be trapped. The first sorts of traps are shallow ones and the second types are deep ones. These traps can be of chemical or physical nature. According to Patsch [15], they may be foreign molecules, carbonyl groups or even polarized states that are able to capture a charge via self-trapping.

On the other hand, the charge concerned may be trapped in different sites associated with amorphous region, amorphous-crystalline boundaries and crystalline region, since the polymer studied is semi-crystalline.

In previous works [1] we have shown that, above 150°C, the shape of isothermal charging current change with temperature reflecting that at low temperature the charge is trapped mainly in the amorphous region but for higher temperature the charge penetrate into the sample and is trapped in the crystalline region [16].

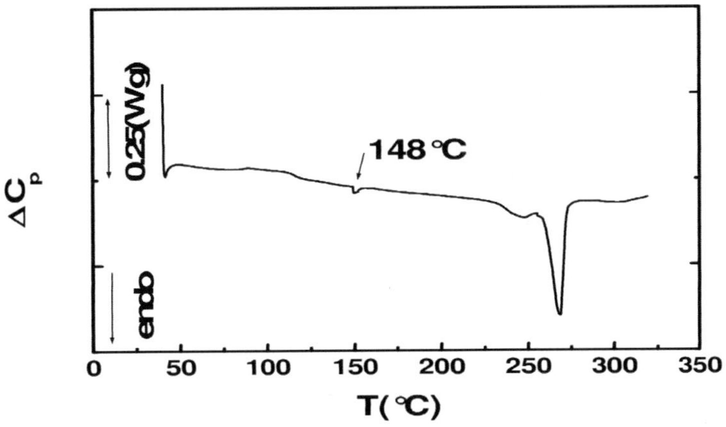

Figure 6. DSC thermogram of semicrystalline PEN.

8. Conclusion

These results implies that, after isothermal discharging there was still space charge trapped which cannot release from the sample and require an increasing of temperature. Indeed, the peak in non-isothermal discharge current at T_g, indicate that the change of polymer state, from glassy to rubbery, favor molecular motion and then facilitate the space charge detrapping process. The second peak was associated to premelting of crystallites.

9. References

[1] Saidi-Amroun N, Oubouchou H, Berdous S, Saidi M and Bendaoud M 2006 *Inter. J. Polym. Anal. Char.* **11** 159

[2] Neagu E R and Marat-Mendes J N 2001 *Jap. J. Appl. Phys.* **40** L1160

[3] Saidi-Amroun N, Saidi M, BerdousS, Oubouchou H and Bendaoud M 2007 *Proc. 9th IEEE Int. Conf. on Solid Dielectrics* 482

[4] Amroun N, Saidi M, Bendaoud M and Hoang T G 1993 *Materials Chemistry and Physics* **33** 168

[5] Adamec V and Calderwood J H 1981 *J. Phys. D.* **14** 1487

[6] Miyairi K and Ieda M 1980 *Jpn. J. Appl. Phys.* **19** 1067

[7] Das Gupta D K et al 1979 *J. Phys. D: Appl. Phys.* **9** 2041

[8] Saidi-Amroun N, Berdous S, Saidi M and Bendaoud M 2004 *Proc.8th IEEE Int. Conf. on Solid Dielectrics* 137

[9] Wintle H J 1974 *J. Non. Crist. Solids* **15** 471

[10] Carmo Lança M, Neagu E R and Marat Mendes J N 2004 *Proc.8th IEEE Int. Conf. on Solid Dielectrics* 205

[11] Neagu E R 2005 J. Appl. Phys. **97** 044103

[12] Gao X, Hou W, Zhou J, Li L and Zhao L 2003 *Macromol. Materials and Engineering* **289** 174

[13] Hardy L, Stevenson I, Boiteux G, Seytre G and Schonhals A 2001 *Polymer* **42** 5679

[14] Saidi-Amroun N et al 2006 *Algerian Journal of Advanced Materials.* **3** 375

[15] Patsch R 1990 *J. Phys. D: Appl. Phys.*, **23** 1497

[16] Neagu E R and Neagu R M 2002 *J. Phys. D: Appl. Phys.* **35** L29

Acknowledgements

We wish to thank the Laboratoire de Génie Electrique (Toulouse) for providing us with sample.

Journal of Physics: Conference Series **183** (2009) 012009 doi:10.1088/1742-6596/183/1/012009

Effect of surface topography and morphology on space charge packets in polyethylene

ZHOU Yuanxiang[1], WANG Yunshan[1], WANG Ninghua[2], SUN Qinghua[1]
[1]State Key Lab of Control and Simulation of Power Systems and Generation Equipments, Dept of Electrical Engineering, Tsinghua University, Beijing 100084, China;
[2]China Electric Power Research Institute, Haidian District, Beijing 100085, China
E-mail: zhou-yx@mail.tsinghua.edu.cn

Abstract. Polyethylene (PE) is a major kind of internal insulating material. With great progresses of space charge measurement technologies in the last three decades, lots of researches are focused on space charge in PE. The heat pressing and annealing condition of polyethylene affect its morphology obviously. During the heat pressing, the surface of PE forms different surface topographies because of different substrate materials. Surface topography has great relation to the epitaxial crystallization layer and influences the space charge characteristic of PE dramatically. This paper studied the formation process of different surface topographies and their micrographic characters in low density polyethylene (LDPE). pulsed electro-acoustic (PEA) method was used to measure the space charge distribution of samples with different surface topographies and morphologies in LDPE. The effect of surface topography and morphology to space charge packet were studied. The surface topography has great influence on space charge packet polarity and morphology has influence on both movement speed rate and polarity of space charge packet.

1. Introduction

Polyethylene is widely used in electrical insulation because of its outstanding insulation performance and mechanical properties. Space charge always plays a very important role in electrical properties of dielectrics. Space charge is well known to distort the local electrical fields, and to affect conduction, breakdown and aging processes [1-3], and thus affect the properties and service life of dielectrics [4]. The space charge packet phenomenon, which appears under higher electrical fields, distorts the local electrical field even more seriously [5-7], and affects the properties of insulating dielectrics more obviously.

The annealing method affects the morphology of polyethylene significantly [8]. During the heat compression moulding process, the surface of the samples forms different epitaxial crystallizations with different substrate materials [9-11]. The epitaxial crystallization affects the space charge properties remarkably according to the research in this paper.

Epitaxial crystallization is the oriented growth of one material on another substrate material, it is a kind of Surface-Induced oriented crystallization. The research of polymer epitaxial crystallization on inorganic salts started during the 1950s, and the epitaxial crystallization between polymers is a new area started during the 1980s [9]. There is research which proves that the cooling rate affects the epitaxial crystallization [12].

Journal of Physics: Conference Series **183** (2009) 012009 doi:10.1088/1742-6596/183/1/012009

It is proved that the epitaxial crystallization affect the mechanical properties of materials [9, 10, 13]. In this paper, the low-density polyethylene (LDPE) was hot pressed with different cooling rate on two different substrate materials to get samples with different epitaxial crystallizations and morphologies, and then observed the space charge packet movement in different sample to research the effect of epitaxial crystallization and morphology on space charge packet.

2. Samples and experiments

2.1. Samples

The raw material used in this paper was pellet LDPE. Its characteristics are as follows: density 0.92 g/cm3, melt index 4.0 g/10 min. Pellet samples were laminated between two pieces of Polytetrafluoroethylene (PTFE) or two glass plates in an oven preheated at 180 °C. They were pressed using a 5 kg steel block, and kept at this temperature for 30 min so as to make the samples melt sufficiently. Then LDPE samples were cooled at a low rate (0.03 °C /s, cooled with hot steel block together), a normal rate (10 °C/s, cooled in air) and a high rate (50 °C/s, cooled in ice water), respectively. The samples are about 100 μm in thickness and about 4 mm in diameter. The samples pressed with PTFE were named as LDPE-F and the samples pressed with glass plates were named LDPE-G.

2.2. The morphology of LDPE

Table 1 shows the crystallinity of the three kinds of samples calculated by FTIR spectrophotometry. The results indicate that different cooling rates result in different morphology in LDPE: lower cooling rate results in larger spherulite and higher crystallinity; and higher cooling rate results in smaller spherulite and lower crystallinity [8]. However, the different substrate materials do not affect the size of spherulite and the crystallinity.

The degree of crystallinity X in % can be obtained by applying the following relation to the infrared spectrum [14]:

$$X(\%) = \frac{337}{(A_a / A_c) + 3.37} \qquad (1)$$

Where A_a is the absorption of the amorphous band at 1303 cm^{-1} and A_c is the absorption of the crystalline band at 1901 cm^{-1}.

Table 1. Crystallinity of samples with different cooling rates.

Cooling rate	High rate	Normal rate	Low rate
crystallinity	43.1%	46.2%	48.8%

2.3. SEM of the surface topography

This paper used the SEM to analysis the surface topography. Figure 1 is the SEM pictures of LDPE-F and LDPE-G samples with different cooling rates. LDPE-G samples have smooth surface, and the LDPE-F samples have rough surface which is epitaxial crystallization layer. Polyethylene forms parallel chain epitaxial crystallization on PTFE substrate, this phenomenon is the oriented crystallization of polymers to the substrate materials [10].

2.4. Space charge measurement of LDPE

Space charge in samples was measured by PEA method [15-17]. DC electric fields of -150 kV/mm were applied to the sample and the the space charge packet movement was observed. The high voltage plate electrode is made of semiconductor and the earth electrode is aluminium. The semiconductor material was polyethylene with carbon black. The earth electrode was an aluminium plate with smooth surface [16]. In this paper the space charge distribution was measured every 5 s.

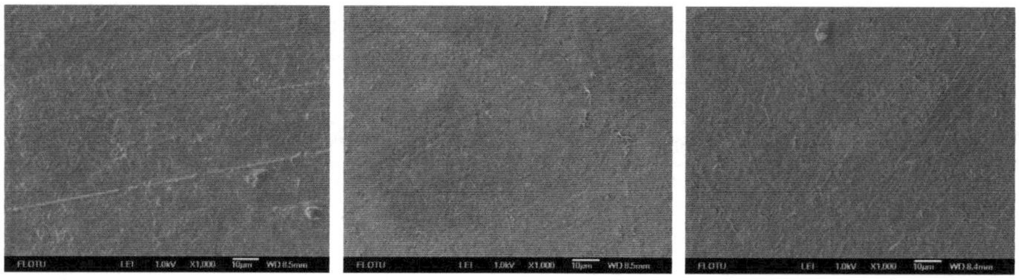

i ii iii

(a)LDPE-F (i: high cooling rate; ii: normal cooling rate; iii: low cooling rate)

i ii iii

(b)LDPE-G (i: high cooling rate; ii: normal cooling rate; iii: low cooling rate)

Figure 1. The SEM photograph of LDPE with different surface topography.

3. Space charge packet characteristic of different morphology and surface topography

3.1. Space charge packet of LDPE-F with different morphology

The LDPE-F samples with surface epitaxial crystallization layer were hot pressed by PTFE substrate material. In this paper a -150 kV/mm were applied to the sample and the space charge packet movement was observed.

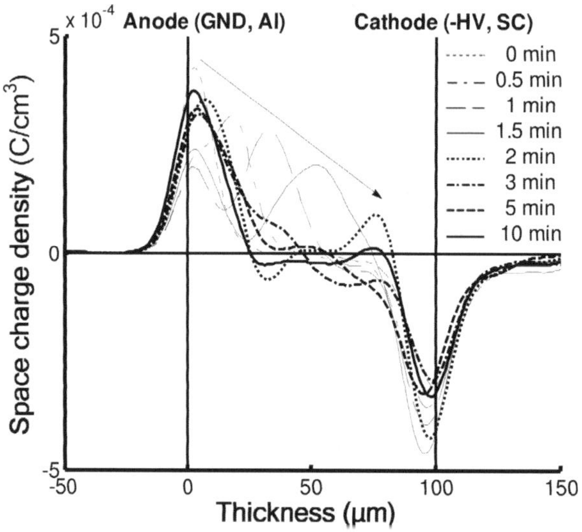

Figure 2. The space charge distribution of high cooling rate LDPE-F sample under a -150 kV/mm field.

Figure 2 shows the space charge movement process in high cooling rate LDPE-F sample. After the voltage was applied, a positive space charge packet formed immediately at the anode side of the sample and then moved toward the cathode side, meanwhile, some negative space charges accumulated at the cathode side of the sample. The positive packet decayed during the movement and finally neutralized with the negative packet at the cathode. A part of the negative charges came out at the back side of the positive packet.

Figure 3 and Figure 4 shows the space charge packet movement process of normal cooling rate and the low cooling rate LDPE-F samples. There ware also obviously space charge packet accumulation and movement as the high cooling rate sample. The polarities of these three kinds of LDPE-F samples were positive mainly.

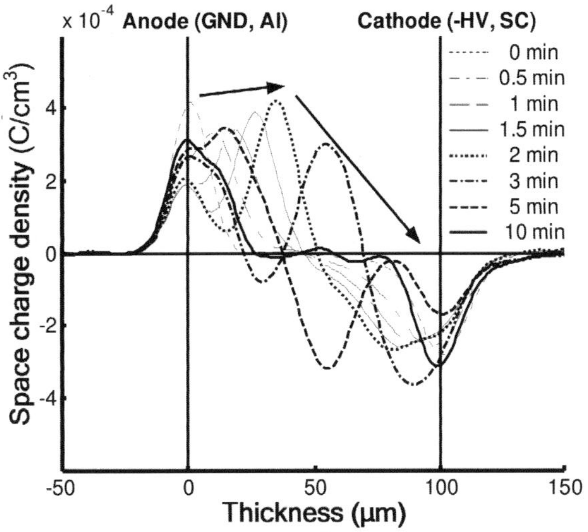

Figure 3. The space charge distribution of normal cooling rate LDPE-F sample under a -150 kV/mm field.

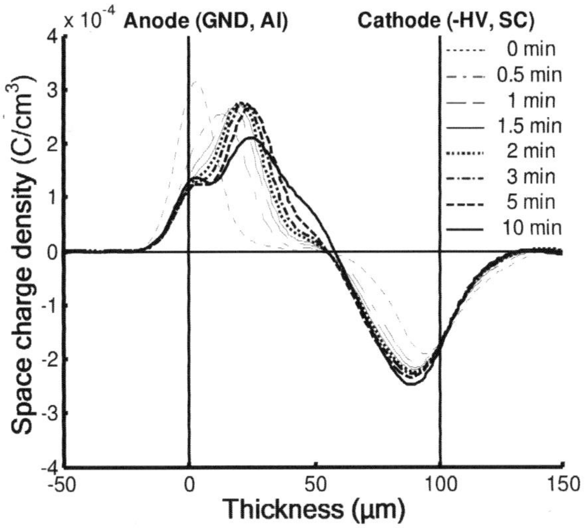

Figure 4. The space charge distribution of low cooling rate LDPE-F sample under a -150 kV/mm field.

The positive space charge packet phenomenon, which forms at the anode side and then moves towards the cathode, was also reported by [18] and [19].

3.2. Space charge packet of LDPE-G with different morphology

The LDPE-G samples without surface epitaxial crystallization layer were hot pressed by glass plates as the substrate material. Figure 5 shows the space charge distribution of high cooling rate LDPE-G sample under a -150 kV/mm field. The positive and negative space charge packets formed at the anode and cathode respectively. The positive packet kept the amplitude at the beginning and then decayed when it moved nearby the negative packet and neutralized with each other. The positive packet decayed to zero when it moved to the cathode. The negative packet moved much slower than the positive packet and did not move near to the anode but stayed in the middle of the sample.

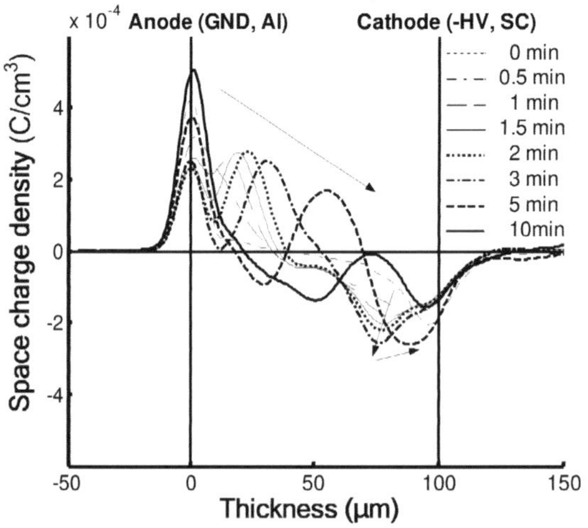

Figure 5. The space charge distribution of high cooling rate LDPE-G sample under a -150 kV/mm field.

During the process described above, the obvious negative space charge packet process was observed. Researches already proved that it is easier to form negative space charge by ionization at the inner side of the cathode without the epitaxial crystallization layer [20]. It accorded with the phenomenon that LDPE-G sample was easier to form negative packet than LDPE-F sample.

Figure 6 and Figure 7 shows the space charge packet movement process of normal cooling rate and the low cooling rate LDPE-G samples. There ware only negative space charge packet appeared in the sample. The negative packet moved towards the anode and the amplitude increased during the movement. The negative packet accumulated at the inner side of the anode finally.

Figure 8 shows the electrical field distribution calculated by the space charge of Figure 7. The negative space charge packet accumulated near the anode and distorted the local field seriously. The field was up to 150% of the applied field. This situation is a severe test to the insulating property of material.

3.3. Discussions of the results

The movement characteristics of positive packet and negative packet have significant differences. The negative packets only appeared in the LDPE-G samples, and the positive packets appeared mainly in LDPE-F samples. The positive packets moved faster than the negative ones and decayed to zero when they moved near to the cathode. The negative packets moved slower, and the amplitude increased during the movement. The negative packets stayed near the anode side of the sample and distorted the local field seriously. The positive packet always repeated several times one after another in a same sample, each packet appeared after the former one decayed to zero. The negative packet only generated once during every experiment. These differences are caused by the morphology and the surface topography.

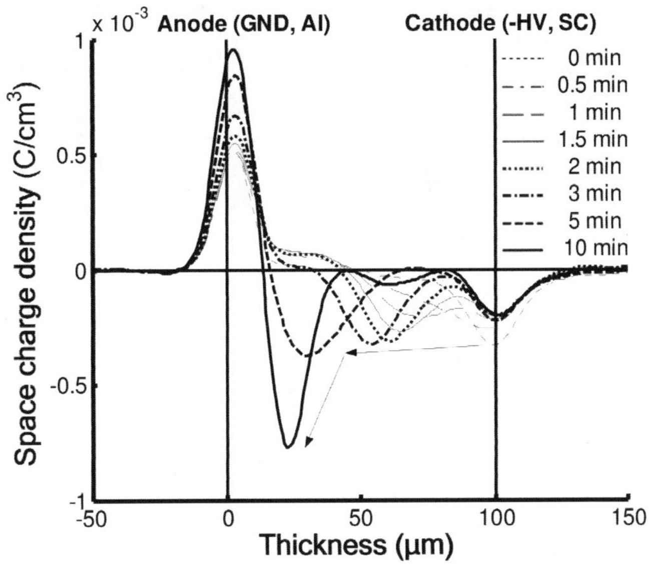

Figure 6. The space charge distribution of normal cooling rate LDPE-G sample under a -150 kV/mm field.

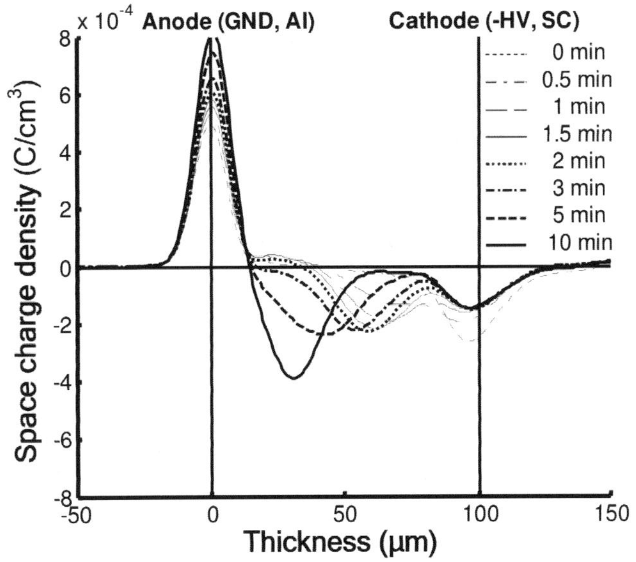

Figure 7. The space charge distribution of low cooling rate LDPE-G sample under a -150 kV/mm field.

The morphology affects the space charge packet phenomenon significantly. The three kinds of LDPE-F samples with different cooling rates show that the space charge packet movement velocity got slower when the cooling rate got slower. When the cooling rate gets slower, the crystallinity will get higher, and the spherulites of the sample will get larger. The spherulites block the transportation of the positive carriers. Some related calculation also showed that the ion mobility decreases while the crystallinity of polymers gets higher [21]. The morphology also affect the polarity of the packets. In LDPE-F samples, when the crystallinity is lower, both of the two kinds of polarity packets exist, however, when the crystallinity goes higher, only positive space charge packet remains.

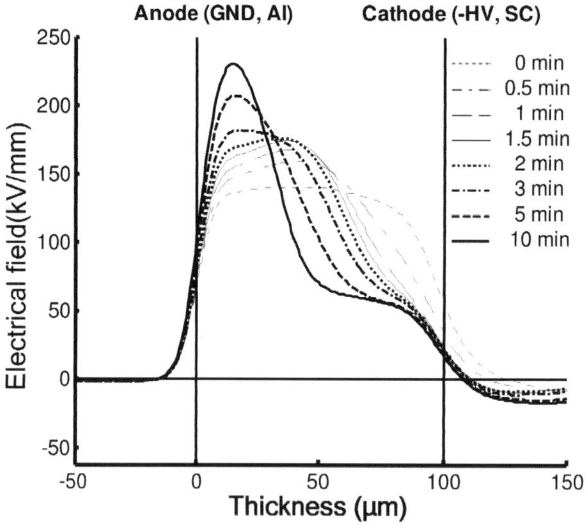

Figure 8. The field distribution of slow –cooling LDPE-G sample under a -150 kV/mm field.

The epitaxial crystallization layer has affection to the space charge packet characteristic. It is comparatively difficult to ionize at the electrode with epitaxial crystallization layer as LDPE-F samples. The space charge packets were form by the positive charge injection of the anode, and the negasitive space charge accumulation was limited. It is easier to ionize at the electrode when the sample have no epitaxial crystallization layer at the surface as the LDPE-G samples. The space charge packets were formed by the negative carriers which were ionized at the cathode.

4. Conclusions

This paper prepared samples with different morphology and surface topography by changing the substrate material and adjusting the cooling rate. The space charge packet characteristics of these samples were studied and the conclusions are as following:

(1) The cooling rate affects the morphology. The higher the cooling rate is, the lower the crystallinity is. The crystallinity is only affected by the cooling rate and has no relationship to the substrate material.

(2) The substrate material affects the epitaxial crystallization layer of LDPE surface. There is significant difference of surface topography between the samples hot pressed by PTFE and glass plates. This is because that an epitaxial crystallization layer generates on the surface of LDPE samples when it is hot pressed by PTFE.

(3) The morphology affects the movement velocity and the polarity of space charge packet. When the crystallinity is higher, the packet movement velocity is lower, and it iseasier to form positive packets in LDPE-F samples while negative packets in LDPE-G samples.

(4) The surface topography has affection to the space charge packet characteristic. The positive space charge packets were form by the positive charge injection at the anode in LDPE-F with epitaxial crystallization layer, while the negative space charge packets were formed by the negative charge ionized at he cathode on LDPE-G without epitaxial crystallization layer.

5. References

[1] Mizutani T 1994 Space charge measurement techniques and space charge in polyethylene *IEEE Trans. on DEI.* **1** 923

[2] Hanley T, Burford R, Fleming R and Barber K 2003 A general review of polymeric insulation for use in HVDC cables *IEEE Elec. Insul. Mag.* **19** 13

[3] Matsui K, Tanaka Y, Fukao T, Takada T and Maeno T 2002 Short-duration Space Charge Observation in LDPE at the Electric Breakdown *Conf. on Electrical Insulation and Dielectric Phenomena.* 2002, Cancun, Mexico
[4] Yin Y, Xiao D M and Tu D M 2002 An Application of Space Charge in Valuing the Electric Ageing Degree of Insulating Polymer *Proc. of the CSEE.* **22** 43
[5] Hozumi N, Suzuki H and Okamoto T 1994 Direct observation of time-dependent space charge profiles in XLPE cable under high electric fields *IEEE Trans. on DEI.* **1**1068
[6] Hozumi N, Takada T, Suzuki H and Okamoto T 1998 Space Charge Behavior in XLPE Cable Insulation under 0.2-1.2 MV/cm DC Fields *IEEE Trans. on DEI.* **5** 82
[7] Zhen F H, Zhang Y W, Guan B and Zhu J W 2005 Formation and Transportation of Space Charge in LDPE *Science in China E.* **35** 701
[8] Zhou Y X, Wang N H, and Yan P 2003 Annealing effect on DC conduction in polyethylene films *J. Electrost.* **57** 381
[9] Mo Z S 1991 Progress in the structure study of crystalline states of polymers *Chinese J. Appl. Chem.* **8** 1
[10] Yin J H and Mo Z S 2001 *Modern Polymer Physics* (Beijing: Scientific)
[11] Yan S K, Lin J and Yang D C 1993 Epitaxial crystallization of layered films of LDPE with highly oriented iPP. *Chinese J. Appl. Chem.* **10** 81
[12] Yan S K, Lin J and Yang D C 1995 Effects of cooling rates on the epitaxy of high-density polyethylene on oriented isotactic polyprolylene substrate films. Acta Polymerica Sinica. **5** 535
[13] He M J, Chen W X and Dong X X 1983 *Polymer Physics* (Shanghai: Fudan University Press)
[14] Teranishi K and Sugahara K 1966 Studies of the Branching and the Crystallinity in Polyethylene by Infraed-Method *Kobunshi Kagaku* **23** 512
[15] Wang N H, Gao B and Zhou Y X 2004 A practical space charge testing system realized by the pulse electro-acoustic method *Elec. Measurement & Instrumentation.* **41** 12
[16] Li Y, Yasuda M and Takada T 1994 Pulsed Electroacoustic Method for Measurement of Charge Accumulation in Solid Dielectrics *IEEE Trans. on DEI.* **1** 188
[17] Takada T, Tanaka Y, Adach N and Qin X 1998 Comparison between the PEA method and the PWP method for space charge measurement in solid dielectrics *IEEE Trans. on DEI.* **5** 944
[18] See A, Dissado A and Fothergill J C 2001 Electric Field Criteria for Charge Packet Formation and Movement in XLPE. *IEEE Trans. on DEI.* **8** 859
[19] Jones J, Llewellyn and Lewis T J 2005 The contribution of Field-Induced Morphological Change to the Electrical Aging and Breakdown of Polyethylene *IEEE Trans. on DEI.* **12** 951
[20] Wang Y S, Zhou Y X and Wang N H 2008 Effect of surface topography on space charge characteristic in polyethylene *China Insul. Mater.* **41** 42
[21] Wang N H 2007 *Morphology Effects on Space Charge Characteristics of Low Density Polyethylene* (Beijing: Tsinghua University)

Acknowledgment
This paper is supported by the Program for New Century Excellent Talents in University (NCET-04-0095), the Research Fund for the Doctoral Program of Higher Education (RFDP 200800030040), as well as the National Natural Science Foundation of China for financial support of the research project NSFC 50277023, NSFC 50347010, NSFC 50436040, NSFC 50877040 and NSFC 50437030.

High mobility conduction in insulating polymers through fast soliton-like charge pulses

D Fabiani[1], G C Montanari[1], L A Dissado[2]
[1]LIMAT Department of Electrical Engineering, University of Bologna, Bologna, Italy
[2]Department of Engineering, University of Leicester, Leicester LE1 7RH, U.K.
E-mail: davide.fabiani@unibo.it

Abstract. We present evidence for the existence of a new form of charge transport in insulating polymers which takes the form of charge pulses that move exceptionally quickly across the insulation in comparison to the charge carrier mobility typical of such materials. This phenomenon could be associated with electromechanical compression of the polymer (a minimum amount of charge is needed for this to occur) that triggers a discontinuous injection of charge at the electrodes and allows transport inside the bulk through chain displacement modes. These pulses behave as charged solitary waves, or solitons, with a speed that corresponds to a mobility of some orders of magnitude higher than that of independent carriers.

1. Introduction

Literature indicates that insulating polymers exhibit generally very low electrical conductivity and the mobility of the charge carriers is low even at high fields (e.g. 10^{-16} -10^{-14} $m^2V^{-1}s^{-1}$ compared to > 10^{-5} $m^2V^{-1}s^{-1}$ for semiconductors and metals) [1-3]. In fact, the electrical conductivity is associated not just with the low carrier mobility, but also with the concentration of intrinsic and extrinsic carriers. Up to quite high fields the carrier concentration is mostly determined by extrinsic carriers as the gap between valence and conduction bands is so high that the concentration of intrinsic carriers is essentially non-existent. Extrinsic carriers are contributed by impurities and contaminants, as well as electronic charges (electrons/holes) injected from the electrode/insulation interface. If the electric field is above a certain value (the so-called "threshold for space charge accumulation" [4]), an excess amount of charge is injected in the material, more than can be extracted from the other electrode, thus accumulating in the bulk as space charge. This charge is trapped in localized states of the material and can move from one electrode to the other by hopping between traps [3]. The crucial feature of this 'traditional' approach is that the carriers move incoherently as point particles independent of one another and without affecting the potential surface on which they move, i.e. the traps are localised and exist whether or not carriers reside in them. In this approach the current is a continuous flow of carriers, which gradually approaches a steady state value after an electric field is applied to the insulating polymer.

Recent experiments however, show that the current picture is not always adequate [5]. Thanks to ultra-fast space charge acquisition systems, small, but repetitive, charge pulses of both polarities, travelling very fast from anode to cathode and *vice-versa*, have been observed in polyethylene specimens at moderate fields. The following sections will provide evidences of this new and unexpected conduction mechanism.

2. Experimental setup

The specimens considered in this paper consist of cable models reproducing HV cables on a reduced-scale (mini-cables), made by three layers: inner semicon (0.7 mm thick), insulation layer of cross-linked-polyethylene (XLPE) material (1.5 mm thick) and outer semicon (0.15 mm thick). The conductor diameter is 2.8 mm (Figure 1).

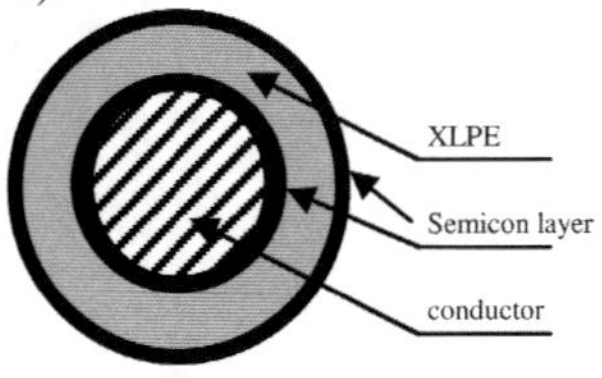

Figure 1. Sketch of tested minicable. Conductor diameter = 2.8 mm; Inner semicon thickness = 0.7 mm; dielectric (XLPE) thickness = 1.5 mm. Outer semicon = 0.15 mm.

Minicable sections of 1.5 m length were kept at 80ºC for 5 days in order to expel volatile by-products produced during the cross-linking of its polyethylene insulation and volatile solvents present in the carbon particle loaded polymer (semicon) that connects the insulating polymer to the metallic electrodes. They were then placed in a controlled temperature environment for the measurement of space charge. This was realised by means of the Pulsed ElectroAcoustic (PEA) technique [6] at field values exceeding the threshold for space charge accumulation, e.g. with average fields of about 40-50 kV/mm (the threshold is around 10 kV/mm at 25 °C). The effect of temperature on charge dynamics was investigated by carrying out tests in an oven under isothermal conditions (no temperature gradient applied to insulation) at 35, 45 and 70 °C.

Space charge profiles were recorded by means of an ultra-fast-acquisition system [5]. The signal to noise ratio is reduced by averaging 100 signals to produce each space charge profile using a digital signal averager that has a fast processor and PC connection. It is therefore able to acquire, average, and store signals at a very high rate (up to 1 MHz). This allows the acquisition and display of space charge profiles every 25ms. The space charge amplitude resolution achieved is about 0.03 C/m^3.

3. Experimental results: the existence of fast charge pulses

A sequence of both positive pulses (from the anode) and negative pulses (from the cathode), see Figures 2a and 2b, are observed contemporarily. The pulse amplitude is small (0.07 to 0.15 Cm^{-3}) and remains around the same value throughout the measurement of the pulse sequence, which occurs with a repetition rate of 2.5 s^{-1} for negative pulses and 3.5 s^{-1} for positive pulses at mean temperature T = 70 ºC and average electric field E = 40kV/mm. Each pulse maintains a constant speed and shape during transit, which takes less than 0.5 s at the range of measurement temperatures and fields. In particular, transit times decrease noticeably with temperature increase from 35°C to 70°C, that is about 3 and 6 times for positive and negative pulses.

The measurements also show that extraction is not immediate. A charged region of opposite polarity to that of the extracting electrode (heterocharge), see Figure 3, is built up as incoming pulses reach the electrode region and are delayed in their extraction. Indeed, even if each charge pulse involves only a small amount of charge (about 0.07-0.1 C/m^3), the pulses are produced with a high repetition rate and therefore after some seconds from the beginning of polarization, several pulses accumulate a noticeable amount of heterocharge close to both electrodes. This heterocharge build up has been observed in our experiments only when a semiconducting layer is present between insulation and metal (copper or aluminium) electrode and for relatively-thick specimens. Tests performed on thin flat specimens having semicon layer only on one side showed, in fact, that no heterocharge was observed close to the electrode not in contact with the semicon [7]. It seems that semicon material can partially block the extraction of injected charge at the counter electrode allowing, thus, a heterocharge build up close to the partially-blocking electrode [7].

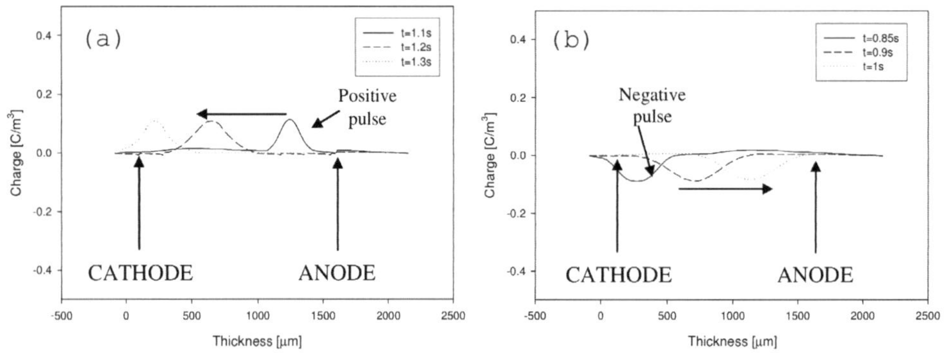

Figure 2. (a) Fast positive charge pulses crossing the insulation, (b) fast negative charge pulses crossing the insulation. The time relevant to each profile acquired is reported in the legend. The arrow indicates the charge pulse direction. Test electric field = 40 kV/mm, temperature = 45°C.

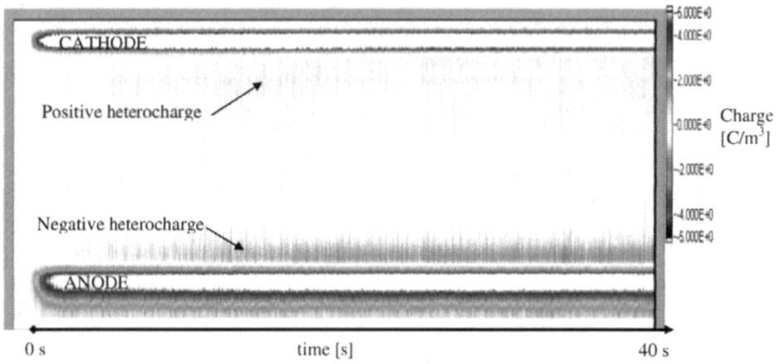

Figure 3. Space charge distribution in the insulation bulk as a function of acquisition time. The colour diagram indicates the accumulated charge amount according to the colour scale reported on the right (cold colours = negative charge, warm colours = positive charge). A heterocharge build up at both electrodes over time is observed. Average field = 50 kV/mm, max polarization time = 40s, temperature = 70°C.

4. Discussion: from charge pulses to solitons

The retention of shape as the pulses cross the sample despite the difference in field between the leading edge and trailing edge (the transit time is too short for carrier diffusion to have an effect) and the lack of acceleration as the counter electrode is approached, implies that such pulses cannot be a cluster of independent carriers. Instead they must be coherent charged entities, i.e. solitons [8, 9]. However the transit time is very fast and our spatial resolution about 40μm, so that we cannot rule out the possibility of a slight redistribution of charge within the pulse; for this reason we have termed the pulse soliton-like [10].

It has to be pointed out that a critical amount of charge, not changing with temperature and field, but depending on the nanofiller concentration, is needed to create a charge soliton. This can be explained by considering the effect of the electromechanical compression at the electrode/insulation interface. An idea to explain soliton injection is displayed in Fig. 4. According to Lewis et al [19] the applied electric field produces a pre-compression of the polymer electrode interface. In addition to this, a given amount of extracharge, ΔQ, which causes an increase of the applied field, ΔE, needs to be available at the electrode/insulation interface in order to produce an extra-compression of the polymer chains against the electrode suitable for the charge ΔQ to tunnel into the chain. After charge injection the interface field and compression drops down, leading the polymer chain to release, thus stopping further charge tunneling. At the end of the process, the charge soliton ΔQ is injected into the material.

It should be observed that this mechanism produces a discontinuous injection of charge pulses, each one having the same size ΔQ. In order to have a repetitive packet injection, in fact, the same critical amount of excess charge at the polymer/electrode interface has to be created to give rise the extra-compression that triggers another packet injection.

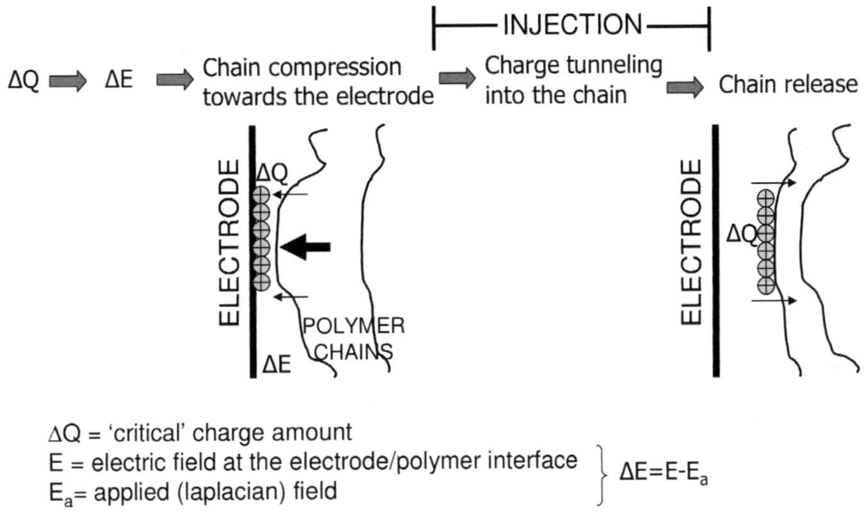

Figure 4. Soliton injection mechanism due to electromechanical compression of the polymer/electrode interface.

In order to provide a quantification of the charge dynamics, the average mobility value, μ, of the soliton-like charge pulses (positive and negative) can be obtained roughly from [1]:

$$\mu = \frac{v}{E} \tag{1}$$

where v is the speed of the charge packet and E is the average electric field applied to the insulation.

The field and temperature dependence of the mobility, Figure 5, allow the main features governing transport of the soliton-like charge pulse to be identified, by use of a general expression valid for hopping and tunneling mechanisms [3, 11],

$$\mu(E,T) = \frac{2kTR}{hE}\exp(-\frac{\Delta U}{kT})\sinh(\frac{eER}{2kT})\exp(-\alpha R) \tag{2}$$

where R is the distance along the field direction the charge carriers in the pulse have to transfer, k the Boltzmann constant, h is Planck's constant, ΔU the thermal activation energy to an excited state from which the particle can tunnel [12] into an electronic state on the receiving site with a probability exp(-αR). The value of the coefficient α is related to the wave function decay and determined by the mass of the tunnelling particle and the barrier through which it tunnels when in the excited state [12].

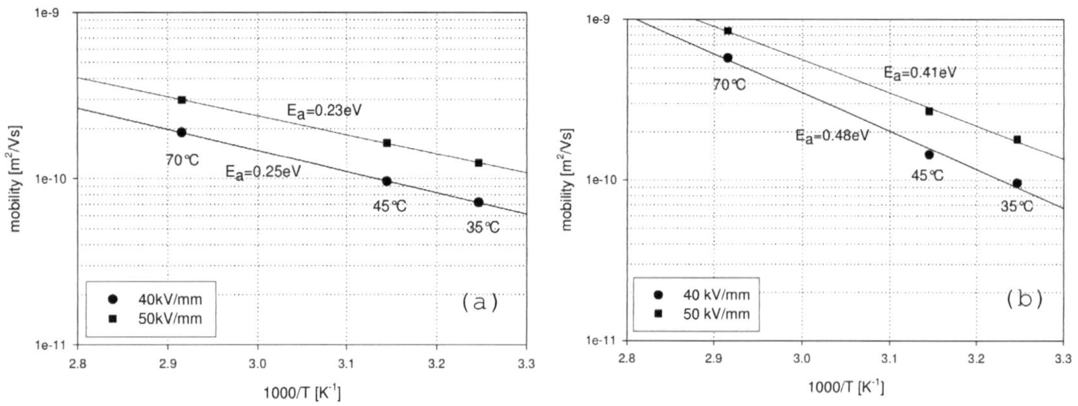

Figure 5. Arrhenius plot of pulse mobility at two different fields. (a) Positive pulses, (b) negative pulses. The overall activation energy, E_a, is also indicated for each field.

In the case of the positive solitons the mobility exhibits an Arrhenius behaviour, i.e. $\mu(E,T)$ is proportional to $\exp(-(\Delta U- eER/2)/kT)$, Figure 5a, with an overall activation energy that is only slightly dependent upon the value of electric field through the term $eER/2kT$ in the exponent. We find that 0.25 eV $\leq \Delta U \leq$ 0.33 eV and 0.05nm $\leq R \leq$ 4nm. It has to be noted that the large range of variation for ΔU and R is due to limits of the experimental accuracy in evaluating the activation energy. The upper bound to R and ΔU has been calculated by assuming the values of E_a reported in Figure 5a, the lower bound has been estimated considering that the field-dependence of the overall activation energy is zero. The tunneling factor, $\exp(-\alpha R)$, can be calculated consequently by inserting the values of mobilities of Figure 4a, ΔU and R in the equation (2). It is found that $1> \exp(-\alpha R) \geq 10^{-3}$.

The field dependence of the overall activation energy of negative soliton mobility is much greater (Figure 5b) and a similar analysis gives $R \approx$ 14nm, $\Delta U \approx$ 0.76eV, and $\exp(-\alpha R) \approx 1$, with a larger accuracy.

These results show that the positive solitons move by a thermally activated tunnelling transfer between sites less than a few nanometres apart. This is consistent with carriers that are positive 'holes' i.e. electron vacancies in the valence bonds of the polymer [11]. Such species can move easily along straight chain sections, but will be trapped at chain bends, twists, and kinks, and will have to transfer between chains by a reverse tunnelling of electrons. Polyethylene and cross-linked polyethylene are semi-crystalline and in extruded insulation such as that investigated will contain crystalline regions in the form of ribbon-like lamellar plates with amorphous regions sandwiched inbetween [2, 11]. Neighbouring chains are not normally close enough to make tunnelling effective, however the so-called crankshaft (or kink) motions [13], operative only in the amorphous regions, can be expected to bring a small region of different chains close enough to allow tunnelling to occur. The activation energy for such motions [14] lies close to the range of ΔU found here. This implies that the positive soliton-pulses are holes bound to kink/crank-shaft distortions in the amorphous regions that travel through the polymer as a solitary wave by inter-chain tunnelling transfer enabled by thermally activated crank-shaft/kink displacements.

The value of $\Delta U \sim 0.76$eV obtained for the negative soliton-pulses would usually be associated with intermediate range electron trap depths [11], which are typical of traps at polymer-free volume interface [16]. However, we note that this value is of the magnitude of the activation energy for the β-motions of the polymer chain [14]. This mode of polymer motion involves whole segments of polymer chain and also operates only in the amorphous part of the polymer. The amorphous regions have a lower density than the polymer crystal and contain potential lattice sites that are unoccupied by atoms of the polymer chain termed free volume. This free volume fluctuates in space and time [17] due to the β-motion displacements. Electrons in polyethylene will trap in the free volume [16, 17] or on the chain bends at the surface of such regions [18]. We suggest that in a negative soliton-pulse, electrons associated with free volume traps drive a field-assisted activation of the β-motion that opens up the polymer chains that inhibit the free transfer of electrons. The electron and the associated free volume are carried together through the amorphous regions as a single entity, a little like the flexing of capillary walls moves blood along in the body. Thus in both cases the soliton-pulses are a quickly moving charged solitary wave in the amorphous region, with the positive and negative soliton-pulses differing in the type of polymer displacement mode with which the charge is bound.

5. Conclusions

Experiments show the existence a new conduction mechanism, i.e. a small amount of charge moves in the amorphous region as solitons, even if the largest concentration of injected charge carriers follows the conventional pattern. The solitary wave nature of the pulse means that there is no inter-change of charge between the two transport systems, which act as effectively independent transport sub-systems. As a consequence, the fast pulsive packets persist into the steady state regime, where they are superimposed on the steady state current as noise fluctuations, small to detect. The fast pulses however dominate in the transient regime. Their build up of heterocharge at the electrode interfaces, which increase the interface fields, may affect negatively insulation system reliability during service.

6. References

[1] Seanor DA *Electrical properties of polymers* 1982 (Academic Press, London, U.K)
[2] Dissado LA and Fothergill JC 1992 *Electrical Degradation and Breakdown in Polymers* (P. Peregrinus for IEE, Stevenage, U.K.)
[3] Mott NF and Gurney RW 1948 *Electronic Processes in Ionic Crystals* (2nd edn Oxford Clarendon, U.K.)
[4] Montanari GC 2000 *IEEE Trans. D.E.I.* **7** 309
[5] Fabiani D, Montanari GC, Dissado LA, Laurent C and Teyssedre G 2009 *IEEE Trans. D.E.I.* **16** 241
[6] Takada T 1999 *IEEE Trans. D.E.I.* **6** 519
[7] Fabiani D, Montanari GC, Laurent C, Teyssedre G, Morshuis PHF, Bodega R, Dissado LA, Campus A, Nilsson UH 2007 *IEEE Electr. Insul. Mag.* **23-6** 11
[8] Stegeman GI and Segev M 1999, *Science* **286** 1518
[9] Bylander J, Duty T and Delsing P 2005 *Nature* **434** 361
[10] Calvo GF, Sturman B, Agulio-Lopez F, Carrascosa M 2002 *Phys. Rev. Lett.* **89** 033902
[11] Jones JP, Llewellyn JP and Lewis TJ 2005 *IEEE Trans. D.E.I.* **12** 951
[12] Razavy M 2003 *Quantum Theory of Tunnelling* (River Edge, NJ, Singapore: World Scientific)
[13] Boyd RH 1985 *Polymer Review* **26** 1123
[14] Crine J-P 1987 *IEEE Trans. E.I.* **22** 169
[15] Mazzanti G, Montanari GC, Palmieri F and Alison J 2003 *J.Appl.Phys.* **94** 5997
[16] Serra S, Tosatti E, Iarlori S, Scandolo S and Santoro G 2000 *Phys. Rev. B* **62** 4389
[17] Artbauer J 1996 *J.Phys.D:Appl.Phys.* **29** 446
[18] Meunier M and Quirke N 2000 *J. Chem. Phys.* **113** 369
[19] Connor P, Jones JP, Llewellyn JP and Lewis TJ 1998 Electric field induced visco-elastic changes in insulating polymer films. *Proc. of IEEE CEIDP*, 27-30

Conductive elastomeric composites

M.H.K.Peace and G.R.Mitchell
Department of Physics, University of Reading, Whiteknights, Reading RG6 6AF UK

E-mail: g.r.mitchell@reading.ac.uk m.h.k.peace@reading.ac.uk

Abstract. Conductive elastic materials are formed by distributing conductive particles within an elastic polymer. We consider a novel composite based on dendritic nickel particles that exhibit remarkably strong negative piezoresistivity with an increase in conductivity of up to 10 orders of magnitude with strains of the order of 0.2. A vital factor for the conductivity of conductive elastomers is the concentration of conductive fillers and many aspects can be understood in terms of percolation theory. In this system the concentration of particles within the composite does not change with strain, yet due to the shape of the particles, the concentration of electrical contacts between the particles does change. We have developed a new model based on the concentration of contact sites, rather than particles which enables us to successfully model this remarkable strain-dependence of conductivity.

1. Introduction

Conductive elastic composites are composites of conductive particles loaded into an elastic matrix in which the conductivity is dominated by the conductive particles [1]. There is a considerable focus in the literature on carbon-based particles, such as carbon blacks, graphite powders and more recently carbon fibres and carbon nano-tubes [2-4]. A variety of phenomena have been observed which include percolation, positive and negative temperature coefficients of resistivity and large strain dependences of conductivity [1,2]. Such electrical measurements have provided insights into the mechanical properties such as the increased stiffness with loading and stress softening [5]. Each of these effects is believed to be determined, to some extent by the formation and breaking of networks of sample spanning chains of inter-connected filler particles [6]. In the case of the effect of strain on carbon black loaded elastomers a number of studies show that there is an initial decrease in conductivity with strain followed by an increase in conductivity with strain > 50% [5-7]. This increase in conductivity is observed for elastic composites with carbon black fillers of high structure, where individual particles are fused aggregates of smaller particles. For composites with lower structure carbon black, the conductivity continually decreases with increasing strain over the whole range of elongation [7]. The strain dependence of conductivity can also be dependent on other factors such as the concentration of conductive particles, the elastic matrix, temperature and method of processing. To our knowledge no detailed mechanism has been presented to explain these variations in conductivity with strain. Percolation theory has been successfully used to explain the change in conductivity of conductive composites with increasing concentration of conductive particles [8,9] with a relationship of the form

$$\sigma = k(p - p_c)^n \quad (1)$$

where σ is the conductivity, p the fraction of conductive particles, p_c the critical concentration, n the critical exponent and k a constant. For conductive composites, p_c depends on the structure and distribution of the conductive particles in the matrix [9] and n is often found to be ~2 [10].

In this paper, results for a conductive elastomer composite with highly structured metal filler particles are presented and we demonstrate the striking negative piezo –resistance displayed by this material. We develop a new approach which applies percolation theory to a system with a fixed concentration of conductive particles but with a strain related concentration of conductive contacts between particles.

2. Materials and Procedures

2.1. Materials

Conductive composites were prepared with poly(dimethyl siloxane) (Unibond) as the elastic host matrix. Dendritic nickel particles (Johnson Mathey) as shown in Figure 1 were used as the conductive filler. The composites were prepared by manually mixing the filler into the matrix polymer until evenly distributed through the mix. Samples for testing were cast into an open topped mould and left to cure for 24 hours. Sample strips 1mm x 8mm x 60mm were cut from the cross-linked sheet.

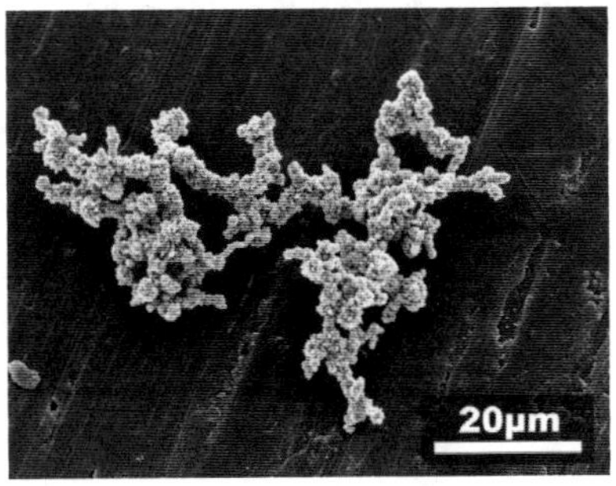

Figure 1. Scanning electron microscope image of the dendritic nickel used in this work.

2.2. Experimental Procedures

A computer controlled tensiometer was used for simultaneously recording the stress-strain data and the conductivity values. Samples were mounted in special clamps that allow four-probe conductivity measurements to be simultaneously taken of the samples as they are being deformed. The strain rate was ~5×10^{-3} s^{-1}. All measurements were made at room temperature.

3. Results

A series of samples were prepared containing different proportions of the filler particles. It was found that the conductivity of composites containing 10% vol., or more of the nickel particles increases dramatically once a threshold strain is exceeded. An example is shown in Figure 2 for a composite containing 20% w/w of nickel particles having an unstrained conductivity of $(6.8 \pm 0.14) \times 10^{-11}$ S/cm. After a strain of 0.045 ± 0.005 the conductivity increases by 10 orders of magnitude over a strain range of 0.2. At higher strains, the conductivity tends towards a plateau value. This data has some similarity to the conductivity-concentration plots of percolating systems (Equation 1), however here the concentration of the conductive filler remains constant. Hence the dramatic rise in conductivity typical of the formation of percolative pathways is not due to a changing concentration. Here we propose that the increase in conductivity is due to an increase in electrical contact between particles. When strain is applied, the geometric centres of the particles move apart from each other according to the deformation of the elastomer. Particle protrusions and extremities of particles that are close

Journal of Physics: Conference Series **183** (2009) 012011 doi:10.1088/1742-6596/183/1/012011

enough to be entangled have a chance of becoming in contact as the separation increases with strain. As strain increases, there is an increasing number of contacts, and decreasing resistance of existing contacts, between neighbouring particles and eventually percolative pathways are formed.

It is well established that metal particles usually contain a thin oxide coating which is non-conducting. Indeed, the loose dendritic nickel powder was found to be non-conducting in the absence of applied pressure[12]. One possible explanation of the threshold strain observed in Figure 2 is that such a strain is required to produce sufficient contact pressure to break the oxide layer allowing conductive pathways to be formed. To test this, we coated the dendritic particles with gold as gold only develops a very thin oxide layer on its surface and hence any contact between Au coated particles will be a conducting contact [11]. The loose gold coated particles were highly conducting in the absence of applied pressure[12]. The resultant strain behaviour of 20% gold coated particles in poly(dimethyl siloxane) is also shown in Figure 2. In the relaxed state this sample has a measurable conductivity.. This suggests that at 20% loading of gold coated dendritic nickel particles, there is a percolating network of contacts. In the case of the non-coated particles, however we propose that the contact forces are insufficient to create conducting contacts in the relaxed state. The rapid rise in conductivity at low strains and a plateau level of conductivity at high strains is still present in the 20% gold coated particle sample. This strongly suggests the formation of new contacts with strain that leads to an increasing conductivity with strain. We have developed a quantitative model based on the concept of contact formation which is presented in the next section.

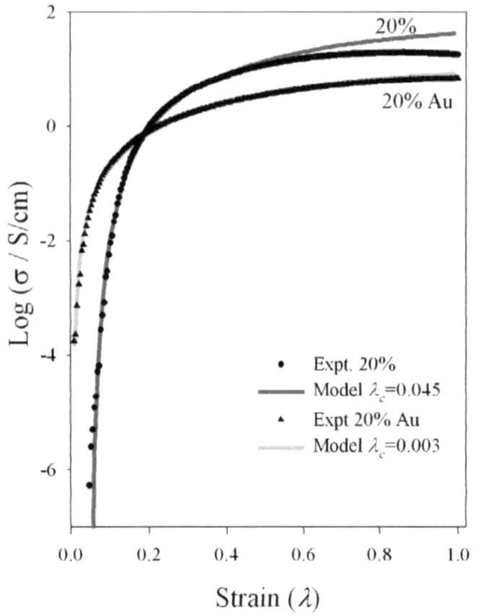

Figure 2. A plot of the conductivity recorded as a function of the strain applied during loading for a conductive composite based on (a) 20% w/w of dendritic nickel and (b) 20% w/w of gold coated dendritic nickel both with a matrix of poly(dimethyl siloxane). The continuous lines represent the predictions of the model developed in this work.

4. Model

Dendritic particles have great irregularity of shape. The protrusions and surface distortions lead to a large potential for entanglement and increased contacts. We use the term overlapping to describe these effects. We consider the overlapping by representing the dendritic-particles as spheres of radius r, where r is the distance from the geometric centre of a dendritic particle. The probability of a volume with a sphere being occupied by the dendritic particle is taken as constant, each particle is rigid and non-deformable and the centres of the spheres move in an affine manner with respect to the deformation of the composite. Any pair of spheres with their centre points located within a distance of 2r will overlap. By considering the geometry of the volume of the overlap we can construct a relationship which describes the concentration of contacts as a function of strain [12]. The particles are not in electrical contact for zero strain and so we introduce a condition that overlapping regions must have a relative deformation greater than a threshold strain before contacts develop. To compare this model with the experimental data, it is necessary to determine how the concentration of contacts relates to the conductivity and for this we use Equation 1, where p is now taken as the concentration of contacts. In this simple model we consider that all particles behave in the same manner and thus as soon as a non-zero fraction of contacts is formed we can assume a sample-spanning cluster of contacts between particles and consequently an increase in conduction; as a consequence the critical concentration of contacts (p_c in Equation 1) is taken as zero. We have taken the exponent in Equation 1 to be 2, the so-called universal value and the scaling factor is the only adjustable parameter.

We compare the predictions between the model and the experimental data in Figure 2 and there is excellent agreement in terms of the basic behaviour. At higher strain there is a divergence between experiment and model. The model outlined here focuses on forming contacts, while the data suggests the breaking of contacts becomes significant at the highest strains. This model has clear implications for other elastomer composite systems. In systems with highly structured carbon black, this model and the inter-locking mechanism could account for the increase in conductivity with strain at intermediate and higher strains. It is possible that at lower strains, the breaking of established contacts with separation of particles dominates, with the inter-locking mechanism operating at all strains, but dominating at higher strains and causing the commonly observed inversion point as discussed earlier. When this new model is applied to spherical particles, there is no degree of overlap of particles and therefore no increase in contact intensity with strain, as observed , for example [12].

5. Summary

A deformable composite made with dendritic nickel particles and poly(dimethyl silxoane) shows a dramatic increase in conductivity when strained. This increase in conductivity has been attributed to an increase in electrical inter-particle contacts of the highly structured metal particles with strain. The increase in electrical conductivity with strain arises as a consequence of an increasing number of contacts. A model has been developed to describe how the concentration and conductance of contacts, between dendritic particles in a deformable medium increase with applied strain. We observed good agreement between the model and the experimental data.

6. References

[1] Norman R.H., Conductive Rubbers and Plastics, Elsevier Publishing Company Ltd. (1970)

[2] Sichel E.K., Carbon Black-polymer Composites, The Physics of Electrically Conducting Composites, Marcel Dekker Inc., (1982).

[3] Sandler J.K.W., Kirk J.E., Kinloch I.A., Shaffer M.S.P, Windle A.H., *Polymer*, **44**, 5893-5899, (2003)

[4] De S., Lyons P.E., Sorel S., Doherty E.M., King P.J., et. al., *ACS Nano*, **3**, (3), 714-720 (2009)

[5] Voet A., Morawski J.C., *Rubber Chem. Technol.* **42**, 7675-7777, (1974)

[6] Yamaguchi K., Busfield J.J.C. and Thomas A.G., *J. Polym. Sci. : Part B: Polym. Phys.*, **41**, 2079-2089, (2003)

[7] Voet A., Sircar A.K. and Mullens T.J., Rubber Chem. Technol., **42**, 874-891, (1969)

[8] Moonen J.T., van der Putten D., Brom H.B., Brokken-Zijp J.C.M., Michels M.A.J. *Synthetic Met.*, **41-43**, 969-972, (1991).

[9] Stauffer D., and Aharony A., Introduction to percolation Theory – Second Edition, Taylor and Francis, (1992)

[10] Balberg I., Phys. Rev. B., **57(21)**, 13351-13354, (1998)

[11] Holm R., *Electric Contacts*, Almqvist & Wiksells, (1946)

[12] Peace M.H.K., *PhD Thesis*, University of Reading 2006

Acknowledgement

We acknowledge support for Mark Peace through a University of Reading Postgraduate Studentship.

Journal of Physics: Conference Series **183** (2009) 012012 doi:10.1088/1742-6596/183/1/012012

The study of molecular movements in dielectrics using isothermal and non-isothermal current measurements

E R Neagu[1,2], C J Dias[1], M C Lança[1] and J N Marat-Mendes[1]

[1]Departamento de Ciência dos Materiais, Secção de Materiais Electroactivos (CENIMAT/I3N), Faculdade de Ciências e Tecnologia, Universidade Nova de Lisboa, 2829-516, Caparica, Portugal
[2] Department of Physics, Technical University of Iasi, B-dul D. Mangeron 67, Iasi700050, Romania
E-mail: neagu@dcm.fct.unl.pt

Abstract. The electrical methods used to study the molecular movements in dielectric materials are based on the movement of the dipoles, which are parts of the component molecules, under a convenient external applied *DC* or *AC* electric field. We have proposed recently a combined isothermal and non-isothermal measuring protocol to analyze charge injection/extraction, transport, trapping and de-trapping in low mobility materials. During an electric polarization process or an electric charging process, besides the polarization process, electric charge is injected into the material. This charge is partially transported and partially trapped in the material, depending on the material properties and on the experimental conditions. As a consequence of the charge trapping process, the structure is *decorated* with space charge and during the subsequent thermally stimulated discharge current experiment we are observing an *apparent peak* and *the genuine peaks* that are related to dipole randomization and charge de-trapping determined by the molecular movements. The aim of this paper is to present a combined measuring protocol that allows to separate the apparent peak and the genuine peaks in polar and non-polar dielectric materials. The method is very sensitive, very selective and allows to reveal, in a single run, details that are not observed when other electrical and non-electrical techniques are used.

1. Introduction

For characterization of the dielectric materials and for a better understanding of the connection between their structure and their properties, it is necessary to use more and more sensitive methods to study the molecular movement at low dimensional scale approaching the nanometric scale. Detailed information on molecular motions in polymers can be obtained using the so-called Dynamic Mechanical Analysis [1], Differential Scanning Calorimetry (DSC), *AC* dielectric relaxation technique [2-6], etc. As concerns the electrical methods, there is always a contribution from the movement of electrons and ions, especially at low frequency and/or high temperature. Traditionally, the electrical methods used to study the molecular movements are based on the movements of the dipoles that are parts of the molecules [2-6]. In the case of the Thermally Stimulated Discharge Current method (TSDC) a *DC* field is applied to orientate the dipoles. Even in the case when the Windowing Polarization (WP) method is applied [4,5] the peaks are quite broad and the measured current is determined by the detrapped space charge (SC) and by dipole randomization. In our opinion the peaks are quite broad because during the charging step too much charge is injected into the sample. To overcome this difficulty we have proposed a combined protocol to analyze charge injection/extraction transport, trapping and detrapping in

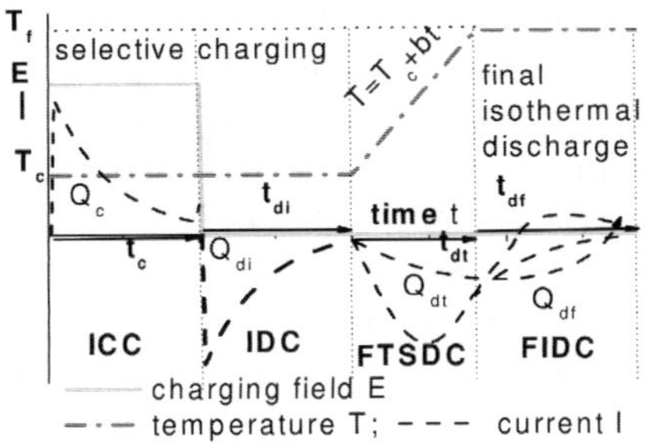

Figure 1. Schematic of the measurement protocol for the employed technique: variation of temperature, electric field and electric current with time during the four different steps.

low-mobility materials [7-9]. The combined protocol consists of four measuring steps: the isothermal charging current (ICC), the isothermal discharging current (IDC), the final thermally stimulated discharge current (FTSDC) and the final isothermal discharging current (FIDC) technique. The experimental conditions can be chosen so that the FTSDC and the FIDC are mainly determined by the movement of detrapped SC.

The aim of this article is to reveal the use of the FTSDC technique as a method to investigate the molecular dynamics in polar and non-polar materials. The method allows to separate the *apparent peak* from the *genuine peaks* related to polymer chain vibrations and structural changes and to reveal fine details not always observed by other methods.

2. Experimental

2.1. Measuring protocol

Figure 1 presents schematically the measuring protocol, giving the evolution with time t of the electric field E, the temperature T and the electric current I during the ICC, IDC, FTSDC and the FIDC measurements. T_c represents the initial temperature at which the sample is charged and at which the ICC and the IDC are measured. T_f represents the final heating temperature and the temperature at which the FIDC is measured. During the ICC measurement a step voltage is applied for a time t_c and the isothermal charging current is measured. At $t = t_c$ the applied DC voltage is removed and the current released from the sample is isothermally measured during the IDC step for a time t_{di}. During this step it is expected that the polarization charge vanishes almost completely so that the IDC gives information primarily about the rapid polarization mechanisms related to the local movement of the molecular dipoles and/or the induced dipoles determined by SC polarization effects at the electrode-sample interface or in the bulk. The time t_{di} must be equal or higher than t_c because (i) the driving field is low during discharging current measurement, (ii) the charge trapping time can be long and consequently the current is low.

During the IDC step, for short times, the depolarization of the lowest activation mode (characterized by a lower value for the relaxation time) takes place. Consequently, the measured current density, which is the time derivative of the polarization, is mainly determined by polarization decay. It is very likely that there is a component of the current determined by a spontaneous and field-assisted detrapping of the trapped space charge. In the presence of trapping, the charge decay process is controlled by the trapping time. For longer trapping times, the current decays very slowly and for a long time period. The shape of the ICC and the IDC provide information about charging and discharging mechanisms and trapping/detrapping processes. During the first two steps (the ICC and IDC) a thermally and field-assisted selective charging takes place. The ratio t_{di}/t_c controls actually the energetic and spatial distribution of the trapped charge. At the end of the selective charging the stored charge is trapped at a mean depth into the material below a certain energy level. This trapped charge will be analysed in the next two steps.

In a usual TSDC experiment a non-equilibrium state is frozen. During the heating of the sample the charge (dipolar or interfacial) relaxes and generates a current whose maximum is usually 10 to 25 K above the poling temperature, as observed using the WP [4].

The situation changes when the new proposed technique is used. At the end of selective charging the sample is heated at a linear heating rate in order to analyze the charge trapped in deeper traps. During the heating of the sample a current is measured which gives information about the field-assisted-charged traps which are now mainly detrapped by thermal movement and structural movements. It is obvious that by choosing the experimental conditions, namely the charging field, charging temperature, work function of the electrodes, environmental atmosphere, t_c and t_{di} we can focus on certain states and valuable information about trapping sites and transport mechanism can be obtained [6-9]. The first and more intense contribution (starting from lower *T*) represents the *apparent peak*, because the peak position changes significantly with experimental conditions [10]. The temperature of the maximum current of an apparent peak is dependent on T_c, t_c, t_{di} and on the heating rate.

2.2. Experimental measurements

Measurements, under a pressure of 5×10^{-3} Torr, were carried out on polar poly(vinylidene fluoride) (PVDF) films of 30 μm thickness and semicrystalline non-polar Teflon FET films of 12.5 μm thickness. The samples of 3 cm diameter were provided with vacuum evaporated silver electrodes (2.5 cm diameter). The charging field was lower than 80 MV m^{-1} and the temperature range from 60 to 320 K and from 273 to 440 K, respectively. The electrical current in the circuit was measured with an electrometer (Keithley 6517A). The electrometer was coupled to a PC for data acquisition and analysis. The sample temperature was maintained constant to within 0.05 °C. The linear heating/cooling rate was controlled by a computer. Before measurements, the samples were kept in a desiccator until they were required for use.

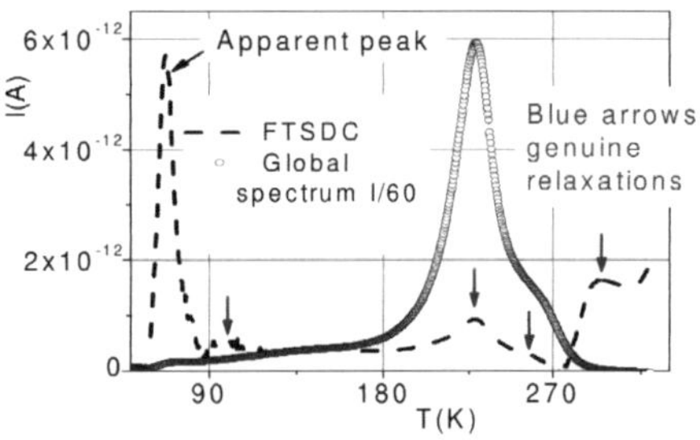

Figure 2. The global spectrum (circles) and the FTSDC for PVDF at low temperature.

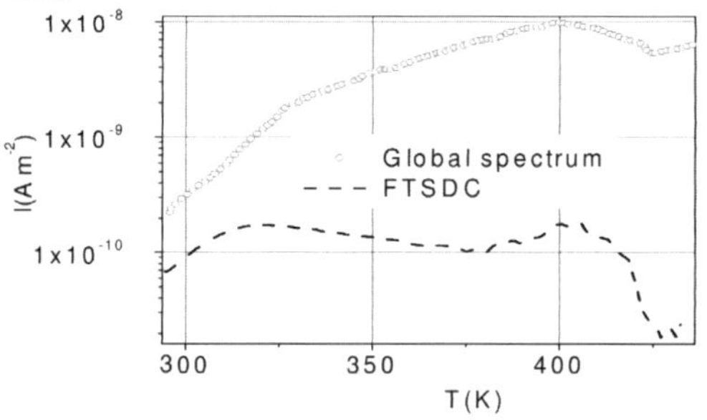

Figure 3. The global spectrum (circles) and the FTSDC for PVDF at high temperature.

3. Experimental results and discussion

Figure 2 displays, by circles, the global spectrum measured for a PVDF sample charged at 273 K for 10 min under a field of 40 MV m^{-1} and cooled under the field to 60 K. Two broad relaxations are observed in the range from 200 to 300 K. The fine structure of the global thermogram in figure 2 can be analyzed using the FTSDC technique. The FTSDC measured for T_c = 90 K and t_c / t_{di} = 5 min/10 min is displayed by a dotted line in figure 2. The first and more intense contribution (starting from lower *T*) represents the *apparent* peak, whose position is around 68 K. Four *genuine* relaxations can be observed around 100, 230, 260 and 290 K. By an appropriate choice of the experimental conditions any of these genuine relaxations can be further analysed.

Figure 3 displays, by circles, the global spectrum measured for a PVDF sample charged at 393 K for 10 min under a field of 10 MV m^{-1} and cooled under the field to 293 K. The broad peak observed around 400 K is the apparent peak. The FTSDC measured for T_c = 293 K and t_c / t_{di} = 5 min/10 min is displayed by a dotted line in figure 3. Two relaxations can be observed around 318 and 400 K. We can conclude that the FTSDC method is more selective and allows observation and separation of the elementary relaxations to a temperature resolution of aproximately 30 K.

A comparison of data in figures 2 and 3 indicates that the method is more selective at low temperatures. This is a consequence of the fact that the sample resistivity is higher at lower temperatures and the charge involved in the charging process is lower.

Figure 4 displays the global spectrum measured for a PVDF sample charged at 320 K for 2 min under a field of 80 MV m^{-1} and four FTSDC thermograms to observe the broad relaxation in the range of 200 to 300 K. The apparent peak becomes broader as T_c gets closer to a genuine relaxation and the genuine peaks are not observed as individual relaxations. The situation becomes similar with the broad peaks observed in *AC* measurements or when the traditional TSDC technique is used. This happens because by increasing T_c more SC is injected and trapped into the sample in deeper superficial and bulk traps. This charge will be released at higher and higher

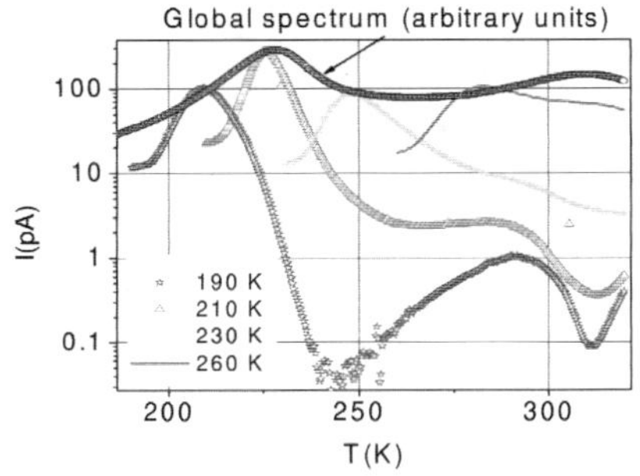

Figure 4. The global spectrum and four FTSDC thermograms to observe the overlap of the genuine peaks as T_c is close to them.

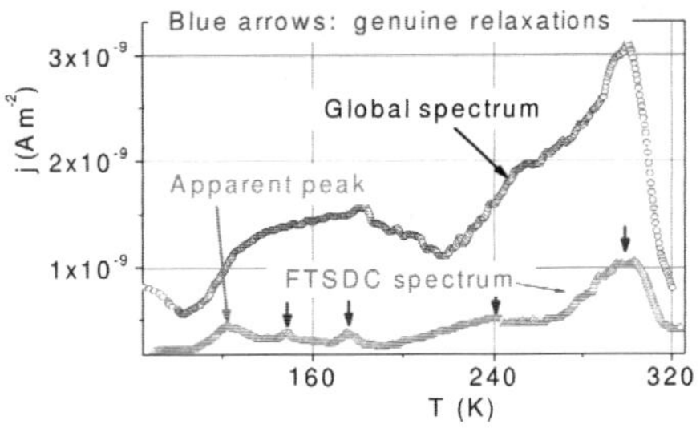

Figure 5. The global spectrum (circles) and the FTSDC for Teflon, a non-polar material, at low temperatures.

temperatures with the consequence that the genuine relaxations will overlap. Consequently, to have a good selectivity and to separate the apparent peak from the genuine peaks or the genuine peaks between them, we have to choose T_c sufficient low in respect with the temperature of a genuine relaxation.

Figure 5 displays, by circles, the global spectrum measured for a Teflon sample charged at 300 K for 10 min under a field of 60 MV m^{-1} and cooled under the field to 60 K. Two broad relaxations are observed in the range from 120 to 320 K. The fine structure of the global thermogram in figure 5 can be analyzed using the FTSDC technique. The FTSDC measured for T_c = 100 K and t_c / t_{di} = 15 min/30 min is displayed by triangles in figure 5. Besides the apparent peak around 120 K, four relaxations around 146, 180, 240 and 300 K can be observed. Because Teflon is a non-polar material we conclude that even the apparent peak is mainly determined by SC detrapping.

4. Conclusions

The Final Thermally Stimulated Discharge Current technique is very selective and very sensitive allowing to separate the apparent peaks from the peaks related to polymer chain vibrations or the structural changes and to reveal fine details not always observed by other methods. It allows to observe and to separate the elementary relaxation in a temperature range of about 30 K. The method can be applied for polar and non-polar materials. The relaxation parameters for local and collective molecular movements can be determined by analyzing the experimental thermograms.

5. References

[1] Menard KP and Sauer BB 2000 *Performance of Plastics* ed W. Brostow (New York: Hanser)
[2] Havriliak S, Jr. and Havriliak SI 1997 *Dielectric and Mechanical Relaxation in Materials* (Munich: Hanser)
[3] Kremer F and Schönhals A 2002 *Broadband Dielectric Spectroscopy* (Berlin: Springer)
[4] Runt JP and Fitzgerald JJ 1997 *Dielectric Spectroscopy of Polymeric Materials* (Washington, DC: American Chemical Society)
[5] Van Turnhout J 1975 *Thermally Stimulated Discharge of Polymer Electrets* (Amsterdam: Elsevier)
[6] Jonscher AK 1996 *Universal Relaxation Law* (London: Chelsea Dielectrics)
[7] Neagu ER and Marat-Mendes JN 2001 *Jpn. J. Appl. Phys.* **40** L810
[8] Marat-Mendes JN, Neagu RM and Neagu ER 2004 *J. Phys. D: Appl. Phys.* **37** 343
[9] Neagu ER and Marat-Mendes JN 2004 *I E E E Transactions Diel. Electrical Insul.* **11** 249
[10] Neagu ER and Neagu RM 2006 *J. Appl. Phys.* **100** (7) art. no. 074107.

Journal of Physics: Conference Series **183** (2009) 012013 doi:10.1088/1742-6596/183/1/012013

Analysis of electrical tree ageing in silicone rubber by physicochemical approach

Y. X. Zhou[1], Q. Nie[1], Z. Z. Chen[2] and R. Liu[1]
1 State Key Lab of Control and Simulation of Power Systems and Generation Equipments, Dept of Electrical Engineering, Tsinghua University, Beijing 100084, China
2 Dept of Electrical Engineering, Zhengzhou University, Zhengzhou 450001, China

E-mail: zhou-yx@tsinghua.edu.cn

Abstract. In this paper, the characteristics of electrical tree ageing in silicone rubber (SIR) under AC voltage were studied. The electrical tree initiation rate is 20% after the application of 6 kV AC voltage for 1000 hours. Samples are separated into three kinds according to processes of electrical tree formation: virgin samples without voltage application, non-treed samples without electrical tree formation after 1000 hours and treed samples with electrical tree formation after 1000 hours. Certain physicochemical diagnostic tests were carried out to understand the degradation of material, ascribed to long-time voltage application, using differential scanning calorimetry (DSC) and thermogravimetric–differential thermal analysis (TG–DTA). Physicochemical analyses, especially the DSC results, show that no additional phases are formed in the processes of electrical tree ageing in SIR. Reduction of the melting point and crystallinity of SIR is observed in the sequence of virgin samples, non-treed samples and treed samples. The activation energy values were calculated from the TG-DTA data. Compared to virgin samples, obvious reduction of activation energy value is observed in non-treed samples. Degradation in SIR has already occurred before electrical tree formation and charge injection and extraction by high field electrode under AC voltage is regarded as the reason.

1. Introduction

With the increase of voltage level, electrical tree ageing is found to be a formidable problem for power cable systems including cable and cable accessories, which are increasingly used in urban areas, hydropower station and pumped-storage power station [1, 2]. Silicone rubber (SIR), as an advanced internal insulating material, has been widely used in high voltage cable accessories due to its excellent insulation and mechanical performance. Although electrical tree phenomena have been found in SIR, enough attentions have not yet been paid. Studies on SIR were mainly concentrated in the external insulation characteristics like flashover characteristics, hydrophobic, weathering aging properties, and so on [3-9]. So it is necessary and urgent to study the electrical tree phenomena of SIR for the stable operation of electrical power system and the development of insulation materials. In this paper, the characteristics of electrical tree ageing in SIR under AC voltage were studied.

The process of electrical tree initiation in SIR is a highly complicated process. Under normal operating voltage, some defect site in SIR, such as gas cavities, conducting inclusions and intrusions, may lead to partial breakdown. Depending on the size and shape of the defect, the stress distribution in the insulation material is altered, causing joule heating/oxidation of the material and a reduction in breakdown strength/local erosion forming prebreakdown channels. In one case, these channels, formed around the defect site in the dielectric structure, resemble branches of a tree and thus the term "electrical tree" is given to the deleterious process. In other cases, although the structure of insulating material around the defect site has already changed, but defect structures that resemble tree branches are not seen and thus the term "electrical tree ageing" should be given to this deterioration process. In this paper, physicochemical approaches are used to analyze the characteristics of electrical tree ageing in SIR to supply an available method to detect the invisible degradation processes in electrical tree ageing.

2. Experimental details

The classical needle-plane electrode was chosen to study the electrical tree ageing characteristics of SIR. The schematic structure and dimension of sample with needle and plane electrodes is shown in Figure 1. The high voltage electrode and ground electrode are all semi-conductive SIR electrode. A stainless steel needle electrode with diameter 250 μm, needle tip 30°, tip curvature radius 2μm, was used to simulate the defect in SIR. The distance between the needle tip and plane electrode is 4±0.1 mm. The raw material used in this paper was two components liquid SIR. After two components of liquid SIR were mixed, needle electrode was put into the liquid unvulcanized rubber. Then the mixed rubber with needle electrode was put into the prefabricated mould and was heated to 100°C to vulcanize. The micrograph of sample with needle electrode is shown in Figure 2.

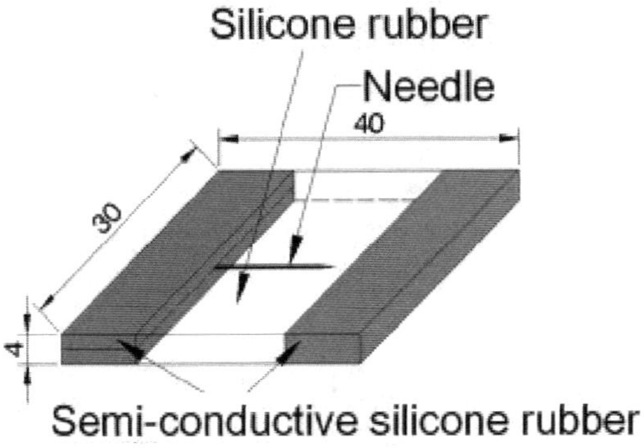

Figure 1. Schematic structure and dimension of sample with needle and plan electrodes.

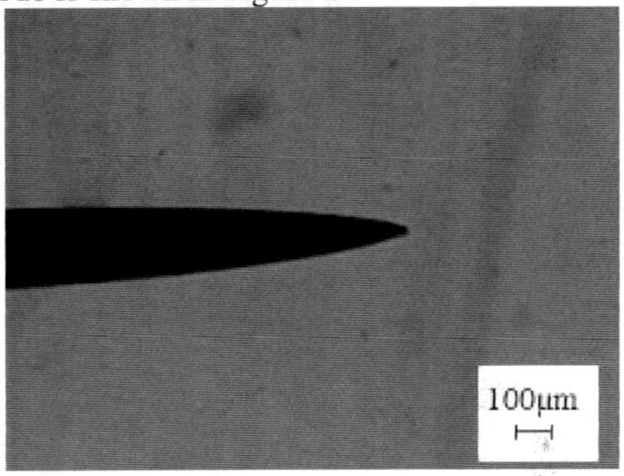

Figure 2. Micrograph of sample with needle electrode.

According to the former studies [10], the mean value of tree initiation voltage at power frequency is 9.7 kV and the minimum value is 7.2 kV. Consequently, 6 kV AC voltage with frequency of 50 Hz was applied on 20 samples to study electrical tree ageing characteristics of SIR. Electrical tree processes were observed by a digital image processing set consisting of a microscope, a digital camera and a computer. In order to increase the efficiency of experiment, a digital micro-observing system for electrical tree experiments of multi samples, as shown in figure 3, were designed and the application number of Chinese patent is 200810238886.8.

The electrical tree initiation rate is 20% after the application of 6 kV AC voltage for 1000 hours. Samples were separated into three kinds according to processes of electrical tree formation: virgin samples without voltage application, non-treed samples without electrical tree formation after 1000 hours and treed samples with electrical tree formation after 1000 hours. Certain physicochemical diagnostic tests were carried out to understand the degradation of material, ascribed to long-time voltage application, using differential scanning calorimetry (DSC) and thermogravimetric–differential thermal analysis (TG–DTA).

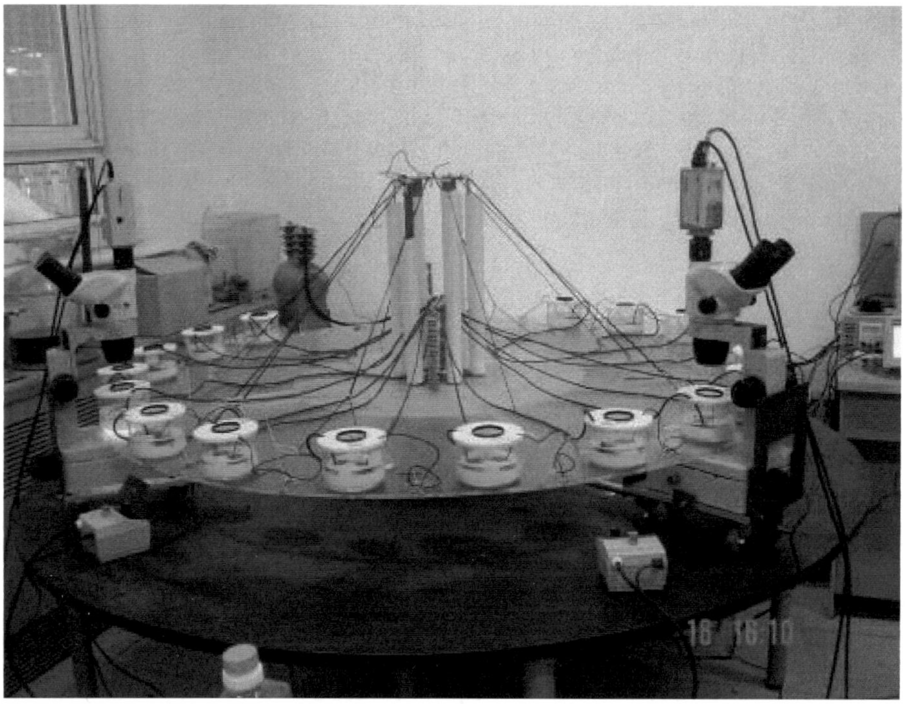

Figure 3. Digital micro-observing system for electrical tree experiments of multi samples (*Application number of China patent: 200810238886.8*).

3. Result and discussions

3.1. Differential scanning calorimetry (DSC)

This technique involves the measurement of energy necessary to establish a zero temperature difference between the specimen and a reference material when the two specimens are subjected to thermal degradation. The melting behaviour of the specimens was observed using a Perkin–Elmer model DSC-2C apparatus (Perkin Elmer Cetus Instruments, Norwalk, CT). The experiments were performed in a nitrogen atmosphere, at a heating rate of 10°C/min. Alumina was used as a standard.

DSC currents of different kinds of silicone rubber samples are shown in Figure 4 and the calculated melting point, caloric content and crystallinity of different kinds of silicone rubber samples are shown in table 1. The DSC results show that no additional phases are formed in the processes of electrical tree ageing in silicone rubber. Reduction of the melting point and crystallinity of silicone rubber is observed in the sequence of virgin samples, non-treed samples and treed samples.

Silicone rubber is a highly micro filled polymer system composed basically of polydimethylsiloxane (PDMS) polymer and 'micro' fumed silica filler. Silicone rubber is composed of the chain of –Si–O– and there are two symmetric methyl groups in the side chain of Si atom. Because of the good solid stereoregularity of the chain and the small steric hindrance of substituent, crystallization can occur in silicone rubber at low temperature. Meanwhile, the aggregation structure of silica and interactions of silica and PDMS has a great influence on the crystallization properties.

Because the silica and PDMS are connected by hydrogen bonding formed by the hydroxyl in the surface of silica and oxygen atom of PDMS, microcrystalline area is formed with silica as nucleation. The increased physical crosslinking points of the filling system reduce the dynamic mobility of molecular chain and the packing density of macromolecular, and then lead to the slowdown of crystallization. As shown in table 1, the melting point of SIR increases in the sequence of treed samples, non-treed samples and virgin samples. That means, as the extent electrical tree damage increases, the melting point moves to a higher temperature

Charge injection and extraction is considered to be an intrinsic factor. When external electrical field was applied to the samples, the electrons in the bulk of SIR or injected from the needle electrode would be accelerated in negative half cycle, and be drawn back in the positive half circle. If the electrons drift under the applied field and the excitation energy after accelerated is large enough, avalanches will occur, and, hydrogen bond of SIR molecular will be cut by collision with carrier, resulting in the decrease of physical crosslinking points. At the same filling quantity, the more combined PDMS the surface of silica adsorbed, the more orderly connections there are and then the higher the crystallinity of SIR is. Consequently, according to the analyses of DSC, it can be concluded that electrical tree ageing leads to the destruction of the chemical connection between silica and PDMS.

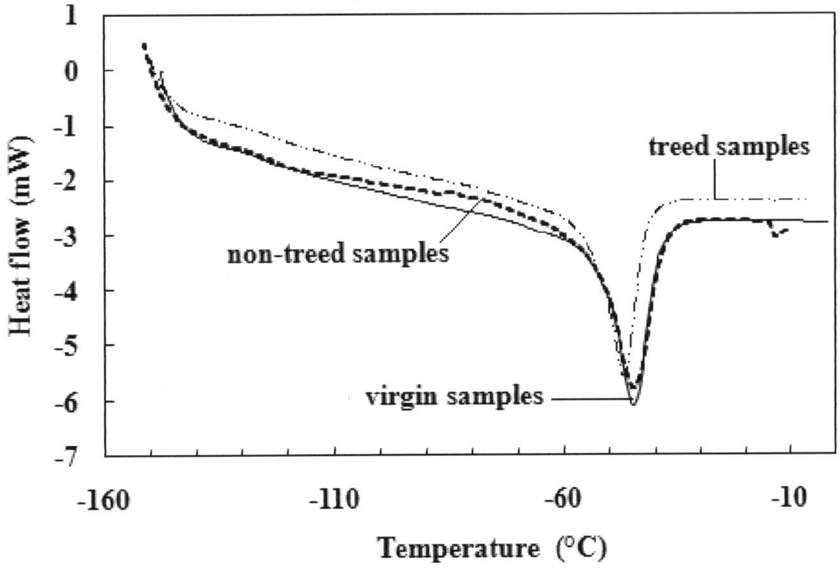

Figure 4. Differential scanning calorimetry currents of different kinds of silicone rubber samples.

Table 1. Melting point, caloric content and crystallinity of different kinds of silicone rubber samples

	Virgin samples	Non-treed samples	Treed samples
melting point	-44.6	-44.8	-47.1
caloric content	24.4	22.1	19.0
Crystallinity	65%	59%	51%

3.2. Thermogravimetric–differential thermal analysis (TG-DTA)

The TG–DTA study was carried out with Netzsch STA 409C equipment. The experiments were performed in a nitrogen atmosphere, in the range 50°C to 900°C. Alumina was used as a standard catalyst. The TG and DTA methods are very effective techniques to study the chemical and physical phenomenon as a function of temperature. Figure 5 shows the thermogravimetric–differential thermal analysis currents of different kinds of silicone rubber samples at the heating rate of 10°C per minute.

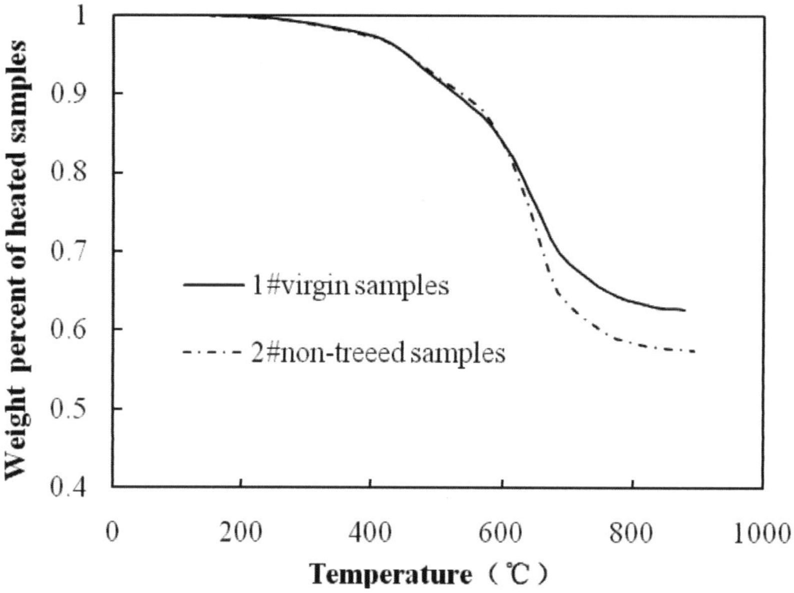

Figure 5. Thermogravimetric–differential thermal analysis currents of different kinds of silicone rubber samples at the heating rate of 10°C per minute.

In order to evaluate the thermal stability of degradation area in SIR further, degradation kinetics of virgin samples and non-treed samples were studied. Samples were heated from 50°C to 900°C in high purity nitrogen protection area with an increase rate of 5°C, 7°C, 10°C, 15°C, 20°C per minute, and their weight change curves were recorded. Because there is only one reaction in the protection of nitrogen, the thermal degradation processes can be described by Arrhenius equation. The rate of weight loss can be described by [11]:

$$\frac{dc}{dt} = (1-c)^n A\exp(-E/RT) \qquad (1)^{[11]}$$

n–reaction order,
A–factor,
E–activation energy,
R–gas constant,
T–Kelvin temperature.

The heating rate is

$$r = \frac{dT}{dt} \qquad (2)$$

Consequently ,

$$\frac{dc}{dt} = r(\frac{dc}{dT}) = (1-c)^n A\exp(-E/RT) \qquad (3)$$

The boundary condition of equation 2 is:

$$(1-c)_{t=0} = 1 \qquad T_{t=0} = T_a \tag{4}$$

T_a is the room temperature. According to the equation of Kissinger [11],

$$\frac{A}{r} = \frac{E}{RT_{\max}^2}\exp(\frac{-E}{RT_{\max}}) \qquad (5)^{[11]}$$

T_{max} is the temperature at which the weight loss rate is the largest. So,

$$\ln(\frac{RT_{\max}^2}{r}) = -\frac{E}{R}\cdot\frac{1}{T_{\max}} + \ln\frac{E}{A} \tag{6}$$

Fitting currents of degradation kinetics of virgin samples and non-treed samples according to Arrhenius equation are shown in Figure 6. The activation energy values were calculated from the TG-DTA data, as shown in Figure 7. Compared to virgin samples, obvious reduction of activation energy value is observed in non-treed samples.

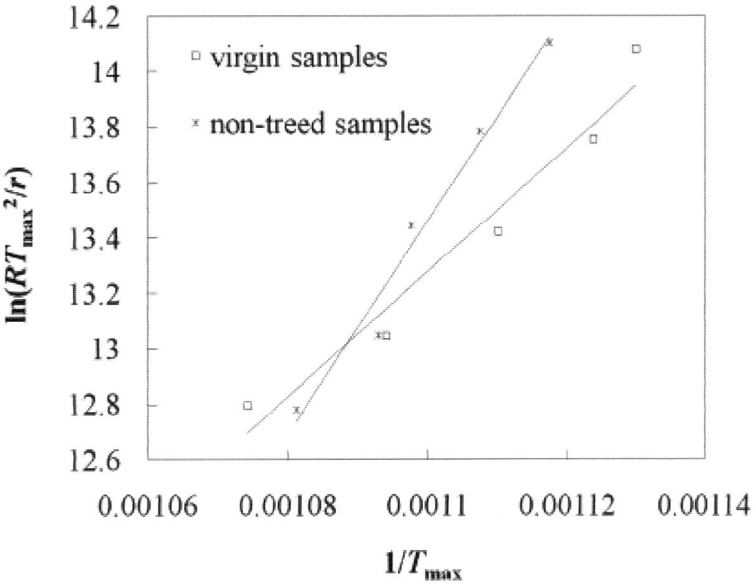

Figure 6. Fitting currents of degradation kinetics of virgin samples and non-treed samples according to Arrhenius equation.

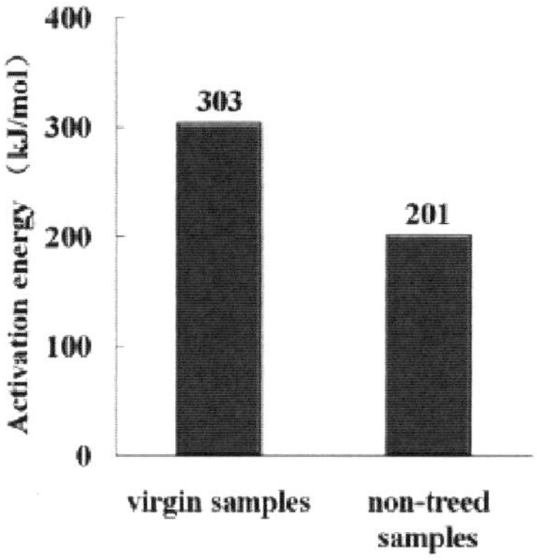

Figure 7. The calculated activation energy values from the TG-DTA data.

4. Conclusion

Physicochemical analyses, especially the DSC results, show that no additional phases are formed in the processes of electrical tree ageing in SIR. Reduction of the melting point and crystallinity of SIR is observed in the sequence of virgin samples, non-treed samples and treed samples. The activation energy values were calculated from the TG-DTA data. Compared to virgin samples, obvious reduction of activation energy value is observed in non-treed samples. Degradation in SIR has already occurred before electrical tree formation and charge injection and extraction by high field electrode under AC voltage is regarded as the reason. Physicochemical approaches supply an available method to detect the invisible degradation processes in electrical tree ageing.

5. References

[1] Tanaka T 1992 *IEEE Transactions on Electrical Insulation* **3** 424

[2] Murata Y , Katakai S and Kanaoka M 1996 *IEEE Transactions on Dielectrics and Electrical Insulation* **3** 361

[3] Liang Xidong , Li Zhenyu and Zhou Yuanxiang 2007 *Proceedings of the CSEE* **27** 19

[4] Li Zhenyu , Liang Xidong and Zhou Yuanxiang 2007 *Proceedings of the CSEE* **24** 30

[5] Liang Ying , Ding Lijian , Li Chengrong , et al 2007 *Proceedings of the CSEE* **21** 7

[6] Sima Wenxia , Liu Zhenyao , Jiang Xingliang , et al 2005 *Proceedings of the CSEE* **6** 113

[7] Lu Zhiwei and Yang Xiuyuan 2001 *Proceedings of the CSEE* **5** 51

[8] Lü Liang , Fang Liang , Wang Xia , et al 2003 *Proceedings of the CSEE* **7** 139

[9] Lü Liang , Wang Xia , He Huaqin , et al 2007 *Proceedings of the CSEE* **15** 106

[10] Nie Qiong , Zhou Yuanxiang , Chen Zhengzheng , et al 2009 *High voltage engineering* **1** 141

[11] Peng ya 2004 *THE PROPERTIES AND REINFORCEMENT OF NANO-CALCIUM CARBONATE FILLED ROOM TEMPERATURE VULCANIZED SILICONE RUBBER* (Sichuan University)

Acknowledgments

This paper is supported by the Program for New Century Excellent Talents in University (NCET-04-0095), the Research Fund for the Doctoral Program of Higher Education (RFDP 200800030040), as well as the National Natural Science Foundation of China for financial support of the research project NSFC 50277023, NSFC 50347010, NSFC 50436040, NSFC 50877040 and NSFC 50437030.

Selection of a suitable vegetable oil for high voltage insulation applications

I L Hosier*, A Guushaa, A S Vaughan and S G Swingler

ECS, University of Southampton, Highfield, Southampton, SO17 1BJ, UK

*E-mail: ILH@ecs.soton.ac.uk

Abstract. Many items of high voltage plant employ a liquid both as a dielectric and a coolant. Currently these systems use a mineral oil, however, this suffers from the drawback of being potentially toxic and hence leakages and eventual disposal can be serious issues. To overcome this problem, an increasing trend in the UK is to backfill existing paper/oil cable systems with dodecylbenzene (DDB). This fluid possesses the advantages of improved gas absorption, good dielectric properties and biodegradability; nevertheless it is still derived from crude oil, a non-renewable resource. Vegetable oils offer the added advantage of being renewable although many types are available with very different properties. In order to select a suitable vegetable oil for high voltage applications, a standardised ageing and testing regime is required. In this paper, a wide range of vegetable oils were subjected to controlled laboratory ageing and the resulting aged oils were characterised by a number of analytical techniques. The results from these tests were then used to rank the different oils, and to select the most ageing resistant oil.

1. Introduction

Mineral oil is widely used as a dielectric medium in high voltage plant but can be damaging to the environment, and hence less hazardous replacements are sought [1]. One solution has been to backfill existing paper/oil cable installations with dodecylbenzene (DDB). Whilst DDB has improved gas absorption properties and is biodegradable, its ageing behaviour is far from ideal [2]. However blends of mineral oil and DDB, which might typically result from backfilling a cable, can nevertheless provide an effective replacement for mineral oil [3]. However, both of these oils are synthetic derivatives of crude oil, a non-renewable resource. A more viable long term solution is to seek alternative oils based on renewable resources. To this aim, vegetable oils have already been applied successfully to small transformers in the United States [4]. It is recognised that vegetable oils have poor oxidation stability compared to mineral oils [5] requiring the use of suitable antioxidants [6]. Whilst vegetable oils are all composed of triglycerides [5], many forms are available each with their own unique properties [7]. In this publication, we have explored the ageing behaviour of six vegetable oils together with a commercial dodecylbenzene (DDB) cable fluid. The oils were aged identically, characterized, and then ranked to determine the most suitable oil for inclusion into high voltage plant.

2. Experimental

2.1. Materials and ageing

Table 1 provides a list of the oils used in this investigation; with the exception of DDB and Envirotemp oil they are all commercially available food grade oils. To determine the oxidation resistance of the different oils, they were each used without further treatment and aged in air. Ageing of 20 ml oil samples was undertaken in fan ovens at 135 °C for periods of up to two weeks. Whilst these ageing conditions are not representative of plant conditions, they do provide a consistent basis for comparison of the different oils over a reasonable timescale. Copper was added to selected oils at a fixed surface area of 12.8 cm^2 to maintain consistency with previous ageing studies [2, 3]. During ageing, the vials were kept covered, to reduce evaporation losses, but were not sealed.

Table 1: Oils used in this investigation

Designation	Description/source
DDB	BICC type C148 batch 5808
Envirotemp	Envirotemp FR3 (Cooper Industries)
Corn	Mazola pure corn oil
Rapeseed	Mazola pure rapeseed oil
Green olive	Filippo Berio extra virgin olive oil
Yellow olive	Asda, light and mild olive oil
Sunflower	Co-op sunflower oil

2.2. Sample characterisation

Ultraviolet/visible (UV/Vis) spectroscopy was performed in a Perkin Elmer Lambda 35 spectrometer using quartz cells of path length 10 mm. Measurements of oil viscosity were undertaken at room temperature using a Physica Rheolab MC1 testing system fitted with a concentric cylinder test cell (diameter 45 mm, length 115 mm, gap 2 mm). Infrared (IR) spectroscopy was performed on a Nicolet 710 FTIR instrument using KBr windows and a path length of 0.1 mm. Dielectric loss measurements were performed at room temperature, using a parallel cup-plate arrangement (diameter 33 mm, thickness 0.1 mm) connected to a Solartron 1296 dielectric interface linked to a Schlumberger SI 1260 impedance gain-phase analyser.

3. Results

3.1. UV/Vis spectroscopy

On ageing, all oils yellow [2-4] and the absorption edge shifts to longer wavelengths (arrowed, Figure 1a). Ageing in the presence of copper causes the ageing to be accelerated; as a result, the oils yellow more quickly and the absorption edge for any given ageing time is shifted further to the right (Figure 1b). Whilst most of the oils displayed a very similar behaviour, several exceptions to this occurred; absorbance peaks associated with carotene (400-500 nm) and chlorophyll (670 and 610 nm) [8] were observed in virgin (un-aged) green olive oil as expected (Figure 1c) but the peaks diminish after

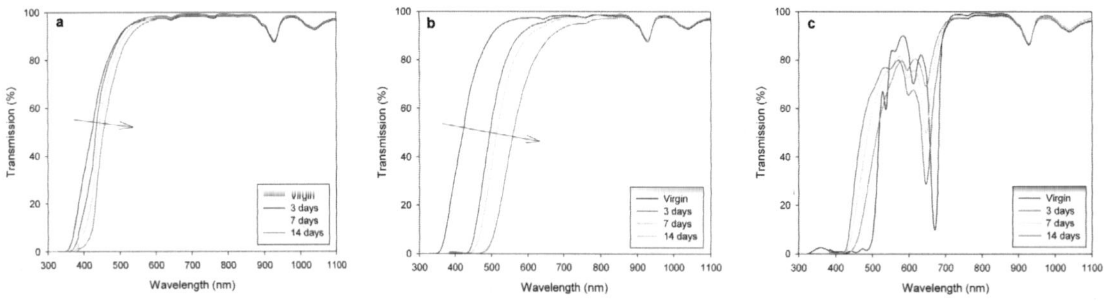

Figure 1: UV/Vis spectra of (a) rapeseed oil aged without copper, (b) rapeseed oil aged with copper, (c) green olive oil aged with copper.

ageing. Similarly, virgin corn oil also contained traces of carotene but no evidence of chlorophyll.

The position of the absorption edge was used, at longer ageing times, where the effects of any carotene present were insignificant, to rank the various oils. The wavelength associated with 50 % transmission is considered here. The oils aged without copper display a very similar behaviour (Figure 2a); the observed variations are close to the typical sample to sample variations inherent to this technique (± 20 nm) so it is difficult to draw firm conclusions. However, ageing with copper (Figure 2b) clearly delineates the differences in ageing behaviour. Olive and Envirotemp oils show the least effects of ageing, rapeseed oil has intermediate properties, whereas corn and sunflower oils both show the greatest change in optical properties after ageing. By contrast, DDB shows the largest change in optical properties after ageing with copper, and the smallest on ageing without copper, and hence shows the widest variation in optical properties of all of the oils. Finally, green olive oil appears to age at a somewhat faster rate than yellow olive oil.

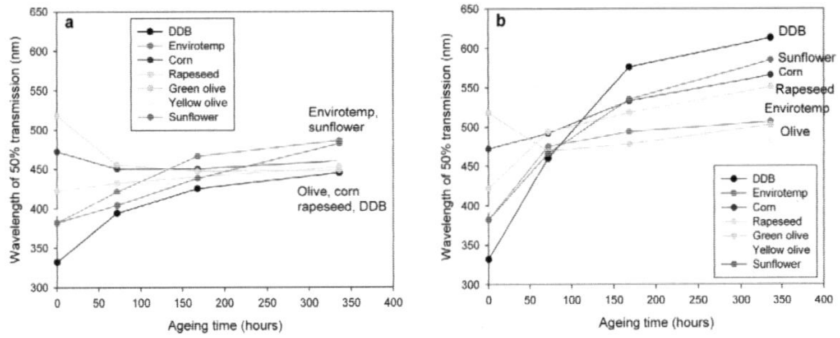

Figure 2: UV/Vis summary plots for oils aged (a) without copper, (b) with copper.

3.2. *Rheometry*

The raw data collected by the instrument was shear stress as a function of shear rate. Since the dependence was linear, the gradients were used to estimate the viscosity of each of the oils. Before ageing, all the vegetable oils had a similar viscosity of ~0.05 Pa s, whereas DDB has a much lower viscosity of ~0.006 Pa s. On ageing without copper (Figure 3a), only sunflower oil shows a significantly increased viscosity. On ageing with copper, more subtle differences are noticeable (Figure 3b). In particular, sunflower and corn oil show the greatest increase in viscosity after ageing, followed by rapeseed oil, olive oil and finally Envirotemp oil. This is the same ranking scheme as deduced above and agrees with available reports [9]. By contrast, DDB shows no significant changes in viscosity after ageing.

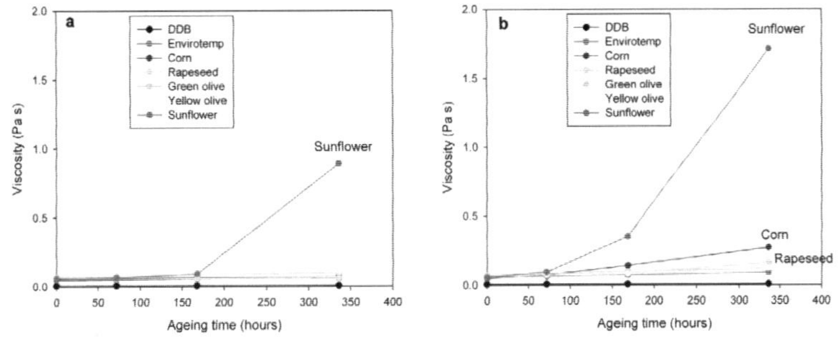

Figure 3: Viscosity of (a) oils aged without copper, (b) oils aged with copper.

3.3. *Infrared spectroscopy*

On ageing without copper, most of the oils showed little or no change in their infrared spectra after ageing, the exception to this was sunflower oil (Figure 4a), where additional absorbance occurs in the hydroxyl region (arrowed) indicating that oxidation is occurring [10]. On ageing with copper, a small but measurable amount of oxidation was observed from envirotemp oil (Figure 4b). This increased significantly in rapeseed oil and finally, sunflower oil displayed significant oxidation, in excess of that shown in Figure 4a, indicating the catalytic effects of copper [2-4]. However, in each case, the same characteristic increase in hydroxyl absorbance occurs indicating the same underlying chemistry [10].

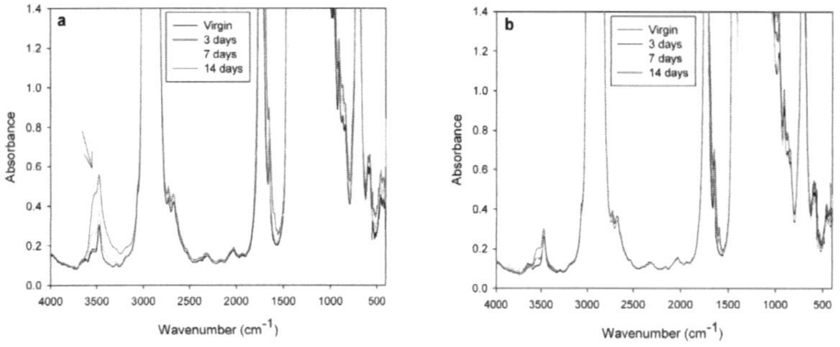

Figure 4: Infrared spectra of (a) sunflower oil aged without copper (b) envirotemp oil aged with copper.

To rank the various oils, the absorbance at a fixed wavenumber (3475 cm^{-1}) was determined. The results confirm that on ageing without copper (Figure 5a) only sunflower oil shows significantly increased oxidation after ageing, as discussed above. Oils aged with copper (Figure 5b) follow broadly the same ranking scheme as that established from the previous two techniques; sunflower oil shows the most dramatic oxidation on ageing, followed by corn oil, then rapeseed oil. Finally, olive and Envirotemp oils both show a small increase in absorbance with ageing time.

Under the current ageing conditions, DDB does not show any detectable oxidation. The use of much larger sample volumes in these experiments, compared to the small volumes used in previous work [2, 3] precludes extensive oxidation from occurring here. Nevertheless, it is clear from the current comparison, where the sample volume and ageing conditions are fixed, that vegetable oils are more susceptible to oxidation than traditional synthetic oils as reported [5, 11].

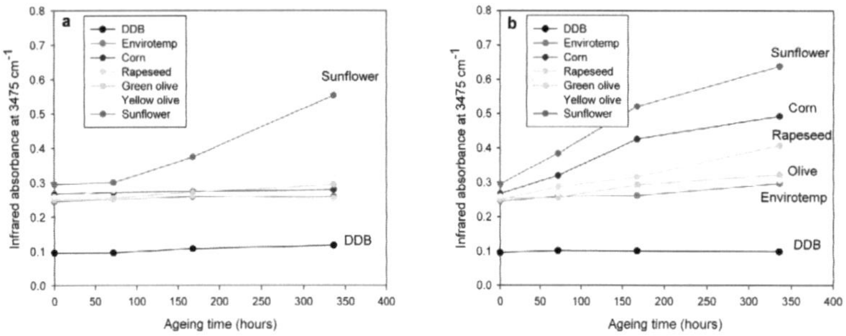

Figure 5: Summary data from oils aged (a) without copper, (b) with copper.

3.4. Dielectric spectroscopy

The dielectric loss increased in all of the oils after ageing [2, 3] therefore, values at 50 Hz were used to rank the oils. After ageing without copper (Figure 6a) the values were very similar for most of the oils although green olive displays the highest dielectric loss even before ageing, and corn oil also displays a significantly higher dielectric loss compared to the remaining oils. This effect suggests that the presence of carotene and chlorophyll serves to increase the dielectric loss.

As noted previously, ageing with copper (Figure 6b) causes a more significant difference between the oils. Whilst the overall ranking scheme of the six vegetable oils largely agrees with that established from the other techniques (i.e. corn oil shows the greatest increase in dielectric loss, rapeseed oil an intermediate increase and olive and Envirotemp oils the least), under that scheme sunflower oil shows a lower than expected dielectric loss and green olive oil shows an increased dielectric loss. DDB shows the lowest dielectric loss on ageing without copper and the highest on ageing with copper, thus it exhibits more extreme changes on ageing, in line with its optical behaviour.

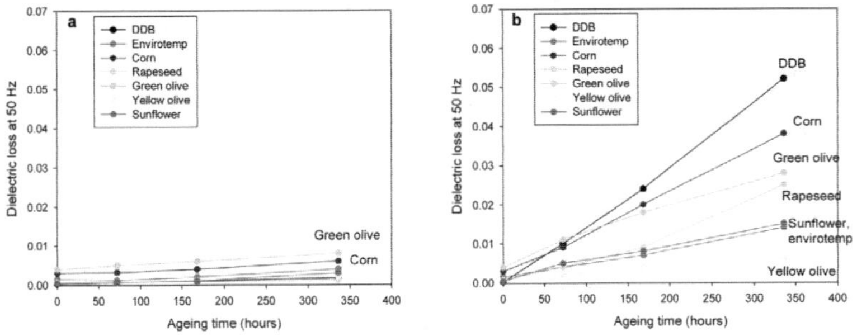

Figure 6: Dielectric loss of, (a) oils aged without copper (b) oils aged with copper.

4. Conclusions

A wide range of vegetable oils were aged, characterised and then ranked according to their ability to withstand thermal ageing in air. Under this scheme, yellow olive oil appears to be the best food grade oil for inclusion into high voltage plant, offering the best resistance to ageing and the lowest dielectric loss. Over most of the indicators of ageing it performed at least as well as the model Envirotemp oil.

Rapeseed oil offers "intermediate" properties and so may find use in some applications, especially if improved through the use of an antioxidant. Oils to avoid are corn oil and sunflower oil, this is mainly due to their tendency to oxidize much more than the other oils and to thicken on ageing. Whilst such thickening could impair oil circulation in equipment, it may also provide a novel route to designing "self sealing" cable systems.

It is clear from this work that carotene and chlorophyll appear to adversely affect the dielectric properties and therefore such materials should be removed from vegetable oils prior to their use in high voltage equipment. Finally, DDB offers significantly improved oxidation resistance compared to vegetable oils in agreement with established literature, however it is particularly susceptible to copper catalysed reactions and hence its dielectric properties are much worse than any of the vegetable oils after identical ageing in the presence of copper.

5. References

[1] McShane C P 2002 *IEEE Ind. Appl. Mag.* **8** 34-41
[2] Hosier I L, Vaughan A S, Sutton S J and Davis F J 2007 *IEEE Trans. Diel. Electr. Insul.* **14** 1113-24
[3] Hosier I L, Vaughan A S and Sutton S J 2007 *Proc. 2007 Conf. Electr. Insul. Diel. Phen.* 69-72
[4] Oommen T V 2002, *IEEE Electr. Insul. Mag.* **18** 6-11
[5] Bertrand Y and Hoang L C 2003 *Proc. 7th Inter. Conf. Prop. Appl. Diel. Mat.* 491-94
[6] Badent R, Hemmer M and Schwab A J 2002 *Proc. 2002 Conf. Electr. Insul. Diel. Phen.* 268-71
[7] Zlatanić A, Lava C, Zhang W and Petrović Z S 2004, *J. Polym. Sci. B: Polym. Phys.* **42** 809-19
[8] White R C, Jones I D, Gibbs E and Butler L S 1977 *J. Agric. Food Chem.*, **25** 143-45
[9] Valdés A F and Garcia A B 2006 *Food Chem.* **98** 214-19
[10] Fox N J and Stachowiak G W 2007 *Tribology Int.* **40** 1035-46
[11] Hosier I L, Vaughan A S, and Swingler S G 2008, *Proc. 2008 Int. Conf. on Diel. Liq.*, 331-34

Acknowledgements

This work was funded through the EPSRC Supergen V, UK Energy Infrastructure (AMPerES) grant in collaboration with UK electricity network operators working under Ofgem's Innovation Funding Incentive scheme; full details can be found on http://www.supergen-amperes.org.

Propylene based systems for high voltage cable insulation applications

I L Hosier[1*], L Cozzarini[2], A S Vaughan[1] and S G Swingler[1]

[1]ECS, University of Southampton, Highfield, Southampton, SO17 1BJ, UK

[2]DMRN, University of Trieste, Via Valerio 6A, 34127 Trieste, Italy

*E-mail: ILH@ecs.soton.ac.uk

Abstract. Crosslinked polyethylene (XLPE) remains the material of choice for extruded high voltage cables, possessing excellent thermo-mechanical and electrical properties. However, it is not easily recyclable posing questions as to its long term sustainability. Whilst both polyethylene and polypropylene are widely recycled and provide excellent dielectric properties, polypropylene has significantly better mechanical integrity at high temperatures than polyethylene. However, while isotactic polypropylene is too stiff at room temperature for incorporation into a cable system, previous studies by the authors have indicated that this limitation can be overcome by using a propylene-ethylene copolymer. Whilst these previous studies considered unrelated systems, the current study aims to quantify the usefulness of a series of related random propylene-ethylene co-polymers and assesses their potential for replacing XLPE.

1. Introduction

Whilst paper/oil cable systems are still commonplace, their use is being gradually phased out in favour of extruded cables employing a solid dielectric medium. The dielectric material of choice is polyethylene, which is cheap, has a very low electrical loss and can be easily processed. However, it must be able to withstand in-service conductor temperatures up to ~90 °C and for this reason, cross-linked polyethylene (XLPE) is widely used due to its enhanced thermo-mechanical stability [1]. However, this material is not easily recycled, posing questions as to its long term sustainability.

Traditional thermoplastics such as polyethylene and polypropylene are currently widely recycled so provide, potentially, a suitable replacement for XLPE in cables over the medium to long term. However, previous studies of ethylene based systems indicate that they lack the required thermo-mechanical stability at high temperatures [2]. In recent years a number of novel new propylene based co-polymers have become available through the use of novel catalyst systems [3] which provide a much more viable option. Propylene based systems have the advantage of offering a much higher melting point than ethylene based systems, making them capable of withstanding high operating temperatures without the need to crosslink. In this regard, previously [4], we have considered a range of unrelated propylene-based co-polymers. We report here on a complementary investigation of a range of related random propylene-ethylene co-polymers and assess their suitability for future recyclable extruded cable systems.

2. Experimental

2.1. Materials and preparation

Five different propylene based systems supplied by Borealis were studied, as detailed in Table 1. HD244 is a standard isotactic polypropylene, SA233 is a random heterophasic propylene-ethylene copolymer whereas the remaining materials are random propylene-ethylene copolymers. Preparation of ~1.7 mm thickness plaques was undertaken using a hydraulic press and aluminium moulds. Preparation of 70 µm films for electrical testing was undertaken using a Grasby Specac 25.011 hydraulic press with constant thickness film maker. All samples were pre-melted at 200 °C for 2 min, pressed, and then allowed to relax for a further 5 min. Samples were then either quenched directly into water at 25 °C or crystallised isothermally between 100 and 125 °C in an oil bath. Crystallisation times ranged between 15 min and 24 h so as to ensure the maximum extent of crystallisation.

Table 1: Polymers used in this investigation

Designation	Ethylene content (%)
HD244	0
RD204	2.2
RD226	3.4
RD208	4.9
SA233	~10

2.2. Sample characterisation

Thermal analysis of ~5 mg samples was undertaken using a Perkin Elmer DSC-7 at a scan rate of 10 K/min. Samples for morphological investigation were microtomed and then etched for a period of 6 hours in a 1 % solution of potassium permanganate dissolved in an acid mixture composed of 2 parts water, 5 parts phosphoric acid and 5 parts sulphuric acid according to published techniques [5]. Samples were then mounted onto standard aluminium SEM stubs, gold coated and examined at 20 kV in a Cambridge Instruments Stereoscan 360 scanning electron microscope. Tensile testing was undertaken at 5 mm/min using an Instron 4301 tensile tester. Dynamic mechanical thermal analysis was performed using a Rheometrics RSA II at a fixed frequency of 100 s^{-1} and a strain of 0.02 % using a dual cantilever geometry. Dielectric breakdown tests were performed according to ASTM D149-87 using opposing 6.25 mm steel ball bearings immersed in a tank of silicone fluid. An increasing 50 Hz AC voltage was applied at a rate of 50 V/s until failure. The results from 20 such tests were then analysed using Weibull statistics.

3. Results

3.1. Thermal analysis

The assignment of the various melting peaks was made with reference to complimentary wide angle X-ray scattering performed on the same samples according to published procedures [6]. A common behaviour with increasing crystallisation temperature was observed. After quenching, a singular melting peak associated with α crystals was always obtained, whilst at intermediate crystallization temperatures, reorganization occurs during the DSC scan, resulting in a "doubling" of this peak. Finally, at the highest crystallization temperatures, an additional lower melting peak, associated with γ crystals, appears which then increases in intensity with further increases in crystallization temperature.

With increasing ethylene content, the latter peak associated with γ crystals becomes more intense and also first appears after crystallization at lower temperatures. In HD244 it first appears after crystallization at 125 °C (arrowed, figure 1a) whereas in RD204 it appears at 120 °C or above (arrowed, figure 1b). In RD226 the same peak (arrowed, figure 1c) first becomes significant after crystallization at 110 °C and finally, in RD208 and SA233 it is apparent after crystallization at 100 °C.

In all these materials, the maximum γ content, obtained after crystallization at 125 °C, ranged from ~5 % in HD244 to ~70 % in SA233. Finally, by considering any fixed crystallization temperature, both the melting peak and enthalpy decrease with increasing ethylene content.

Whilst the details of the crystallization of these materials are clearly complex, the increasing prevalence of the γ phase and the reduction in melting point and enthalpy with increased co-polymer content are in line with similar studies on well characterised propylene-ethylene co-polymers [6]. Most importantly, from the perspective of future cable designs, all the materials considered here display a melting peak in excess of 120 °C, which is significantly higher that the ~105 °C which is normally associated with XLPE [7].

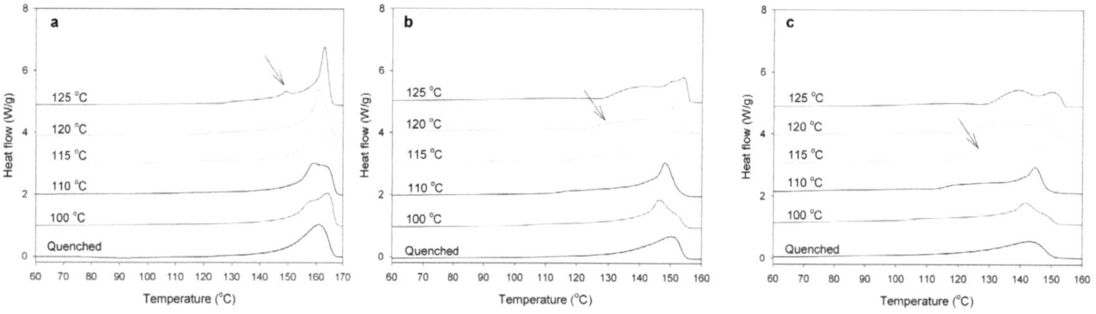

Figure 1: DSC traces from (a) HD244, (b) RD204, (c) RD226

3.2. Morphology

After quenching, small spherulites were observed in HD244 and RD204 as shown in figure 2a, but in co-polymers with > 3 % ethylene content these were replaced with a uniform lamellar texture (figure 2b).

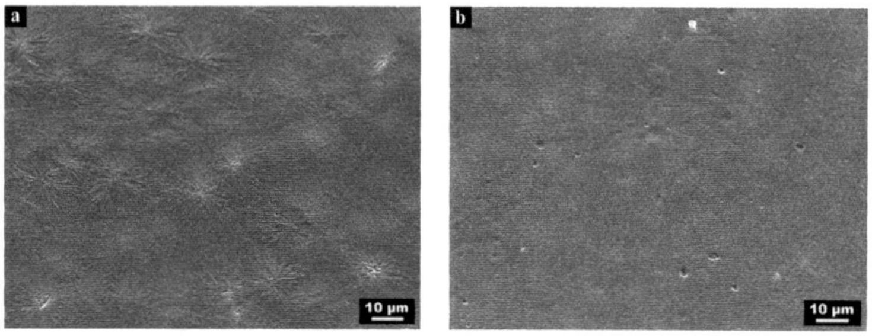

Figure 2: SEM micrographs of quenched samples (a) RD204, (b) RD226

Crystallisation at higher temperatures reduces the nucleation density and so larger spherulites form. In HD244 and RD204 crystallised at the highest temperatures, these possess distinct boundaries (figure 3a). At higher ethylene contents, or at lower crystallisation temperatures, their boundaries become less distinct as shown in figure 3b. Under isothermal crystallisation, spherulitic forms generally persist up to ~5 % ethylene content; in SA233 the morphology is always a uniform texture of fine lamellae. The above effects of crystallisation temperature and ethylene content on morphology are in line with complementary studies on a range of well characterised random metallocene propylene-ethylene co-polymers [8].

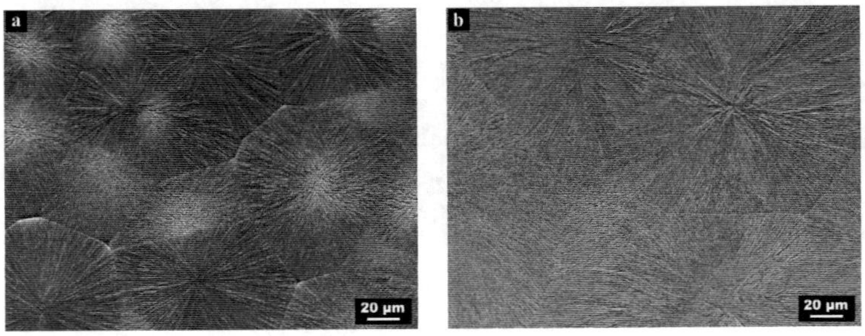

Figure 3: SEM micrographs of isothermally prepared samples (a) HD244 crystallised at 125 °C, (b) RD208 crystallised at 120 °C.

3.3. Tensile testing

After quenching (figure 4a), all five materials display a characteristic necking behaviour [4] after ~20 % imposed strain and do not fail under the loadings imposed here. With increasing isothermal crystallisation temperature, samples low in ethylene content display a much more brittle behaviour. The end result of this is that, at the highest crystallisation temperatures (figure 4b), both HD244 and RD204 catastrophically fail at strains of 15 % and 28 % respectively. These materials both exhibited characteristic brittle fracture surfaces when subsequently examined under an optical microscope. By contrast, the remaining materials, which are higher in ethylene content, retain the necking type behaviour. Whilst none of these materials display the ideal rubbery behaviour associated with XLPE [9], clearly, materials having an ethylene content in excess of ~2 % are required to avoid an unwanted brittle-fracture type of behaviour after crystallising at higher temperatures.

3.4. Dynamic mechanical thermal analysis

The modulus measured at small strains did not show a significant dependence on crystallisation conditions, therefore only data pertaining to quenched samples are shown in figure 4c. The behaviour is in line with the previous propylene co-polymers [4] and HD244 shows, at room temperature, a typical modulus of ~10^9 Pa. This is far greater than the ~10^8 Pa which is typical of XLPE [9]. Nevertheless, with increasing ethylene content, significant reductions in the modulus are achievable, with SA233 showing the most favourable properties. It is clear from the figure that good thermo-mechanical stability is maintained up to ~110 °C without the need for cross-linking.

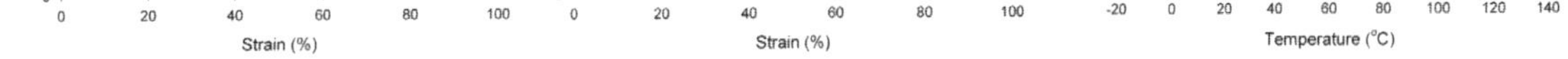

Figure 4: Mechanical test results, (a) tensile tests on quenched samples, (b) tensile tests on samples crystallised isothermally at 125 °C, (c) DMTA results from quenched samples

3.5. Dielectric breakdown

Figure 5 shows breakdown data for various materials. After quenching, the Weibull scale parameter is typically ~180 kV/mm, which is considerably higher than the ~150 kV/mm associated with polyethylene systems [2]. The measured value is independent of ethylene content within the uncertainty associated with these measurements (± 10 kV/mm). However, differences occur between the materials following crystallisation at higher temperatures; in particular HD244 exhibits a significantly poorer breakdown performance when crystallised at 115 or 125 °C (figure 5a) as does

RD204 after crystallisation at 125 °C (figure 5b). The prominent spherulitic boundaries which occur under these crystallisation conditions could explain these results and also the rather poor (brittle) mechanical behaviour of these systems. The remaining materials all exhibit a breakdown strength which is largely independent of crystallisation route, within experimental uncertainties (figure 5c).

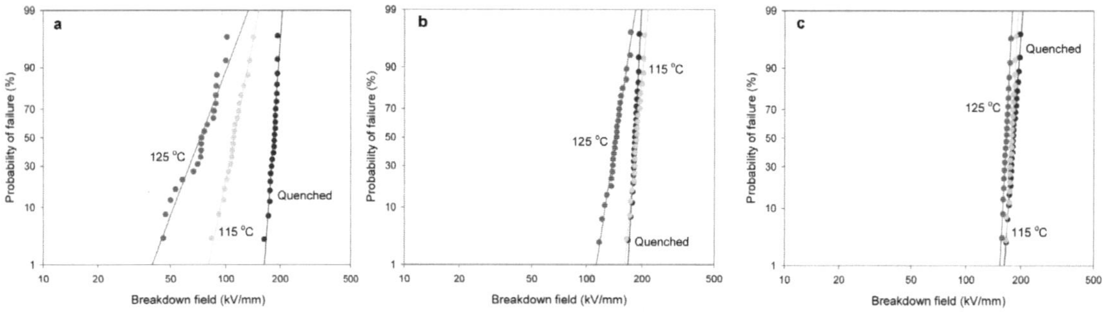

Figure 5: Weibull plots for (a) HD244, (b) RD204, (c) RD226

4. Conclusions

A range of propylene-ethylene co-polymers were examined and the changes in melting behaviour and morphology were in line with previous investigations. Samples low in ethylene content crystallised at higher temperatures were mechanically brittle and had poor electrical strength which could be associated with the rather prominent spherulitic boundaries observed in these cases. A certain amount of ethylene is required to reduce the mechanical modulus, improve the elasticity and ensure that the breakdown strength remains high over all crystallisation conditions. Out of the materials considered here, SA233, a co-polymer containing ~10 % ethylene, offers the most optimal properties for use in a future recyclable extruded high voltage cable.

5. References

[1] Vahedy V 2006 *IEEE Electr. Insul. Mag.* **22** 13-18

[2] Hosier I L, Vaughan A S and Tseng W 2007 *Proc. 9th Intern. Conf. on Solid Diel.* 184 – 7

[3] Langhauser F, Kerth J, Kersting M, Kolle P, Lilge D and Muller P 1994 *Angew. Makromol. Chem.* **223** 151-64

[4] Hosier I L, Reaud S, Vaughan A S and Swingler S G, 2008 *Conf. Rec. of the 2008 Intern. Symp. on Electr. Insul.* 502-5

[5] Olley R H, Bassett D C 1982, *Polymer* **23** 1707-10

[6] Hosier I L, Alamo R G and Lin J S, 2003 *Macromolecules* **36** 5623-36

[7] Hosier I L, Vaughan A S, Campus A and Nilsson U 2007 *Proc. 9th Intern. Conf. on Solid Diel.* 227-30

[8] Hosier I L, Alamo R G, Lin J S 2004 *Polymer* **45** 3441-55

[9] Martin C P 2004 *The impact of mechanical stress on the mechanical integrity of XLPE cables, PhD Thesis*, University of Southampton.

Acknowledgements

We would like to thank Borealis for supplying the materials used in this investigation. This work was funded through the EPSRC Supergen V, UK Energy Infrastructure (AMPerES) grant in collaboration with UK electricity network operators working under Ofgem's Innovation Funding Incentive scheme; full details can be found on http://www.supergen-amperes.org.

Raman microprobe analysis and ageing in dielectrics

N A Freebody[1], A S Vaughan and P L Lewin

School of Electronics and Computer Science, University of Southampton, Highfeild, Southampton SO17 1BJ, UK

E-mail: naf08r@ecs.soton.ac.uk

Abstract. Subsurface voids in samples of electrically stressed low density polyethylene (LDPE) were analysed using confocal Raman microprobe spectroscopy (CRMS). An optical depth profiling technique was used to probe a void along the optic axis whilst a burst void was analysed at various lateral positions. Spectra from the voided samples showed signatures with similar features to those found in previous studies of electrical trees in polyethylene, including the presence of the D and G bands of disordered sp^2 carbon. Results and spectra were then compared to the depth profiles and spectra from block and thin film samples of polyethylene and polystyrene which indicated that the established theory behind CRMS is oversimplified and that the detected signal is largely influenced by the optical properties of the material in question. Overall the study showed that despite some spatial resolution limitations of the technique, depth profiling is a useful tool in the analysis of aged polymers and dielectrics as it can show the variations in chemical composition with respect to position along the lateral and optic axis, a property especially relevant to electrical trees.

1. Introduction

The topic of ageing in dielectric materials has attracted much research and study [1] due to its applications in high voltage insulation. The process of ageing occurs when the material is exposed to intense electric fields. These ageing processes span from the quantum level to complete macroscopic failure. Defects within a material are formed and evolve through partial discharge (PD) activity to form electrical tree like structures.

In recent studies [2, 3] confocal Raman microprobe spectroscopy (CRMS) has begun to reveal the chemical processes involved in electrical treeing and ageing in dielectric materials. This technique is thought to have the potential to provide spectroscopic data with a spatial resolution of the order of ~1 μm along both the lateral and optic axes [4] and, since excitation involves visible photons, the technique is also capable of interrogating the interior of transparent dielectrics, for example polyethylene and polystyrene.

Raman studies of electrical trees in polymers have, however, cast doubt upon these assertions [3] and, if the potential of this approach is to be realised, the true spatial origin of the detected Raman photons must be determined. This paper describes a new approach to this problem and combines optical depth profiling with studies of Raman excitation in a range of materials and sample geometries, including a chemical analysis within voids that have experienced PD activity.

[1] To whom any correspondence should be addressed

2. Materials and Methods

Two samples of low density polyethylene (LDPE) (~700 μm thick) were prepared by heating and pressing a blend of PE and the blowing agent azodicarbonamide into a mould. Heating the samples causes the blowing agent to decompose, hence leaving voids within the sample. One of the samples (Sample 1) has experienced partial discharge within it but has not broken down, leaving a number of closed subsurface voids. In the second sample (Sample 2) of LDPE extensive partial discharges within the material induced a breakdown, causing a void to burst. For comparison to the LDPE samples, block samples (~ 5 mm) were made of polyethylene (PE) and polystyrene (PS) by heating and compressing a mould of the sample. Thin films (~70 μm) of PS were also prepared in the same way.

Samples were analysed using a Renishaw Raman RM1000 microscope using a Renishaw NIR 780TF CW diode laser, wavelength 785 nm and output power of 25 mW, and a x50 objective lens. The spectrometer was set up in confocal mode, with a slit width of 15 μm and a CCD area of 5 pixels (image height) x 574 pixels (spectrometer range), which acts as a virtual confocal pinhole. All spectra of the LDPE samples consist of extended scans between 3500 cm^{-1} and 100 cm^{-1} and were built up of 25 accumulations of 10 s each. In order to minimise sample damage, all spectra were obtained with the laser power set to 25%. The spectra obtained from the PE and PS samples used for comparison were taken over a smaller static range around the main peaks and built up of 10 accumulations of 10s.

3. Results

Initially reference spectra were obtained from the focal surface of the block samples of PE and PS and the main peaks in the spectra identified. The magnitude of these peaks was then recorded as a function of vertical stage position (relative to the focal surface of the sample) hence producing depth profiles of the samples in question. This technique of 'optical sectioning' [5] was also applied to the thin film sample of PS and the depth profiles in figures 1 and 2 were obtained. From these depth profiles in figure 1, we can see that for the samples tested the Raman signal extends both above and below the sample surface. Also once a maximum intensity has been reached the signal declines at varying rates. The signal decline for PE is much steeper than for PS, an effect associated with the optical clarity of each material as this decline is caused by scattering effects within the sample. For the PS film in figure 2, the signal drops off very rapidly due to the lower surface of the sample. The width of the depth profile for the film is however close to half the physical thickness of the sample, which, as Everall discussed in 2000 [6], is due to refraction effects at the upper surface. It is apparent however from figure 2 that it is possible to use CRMS in the analysis of thin films and, hence, the technique can be extended to layered samples, voids and electrical trees in polymeric dielectrics.

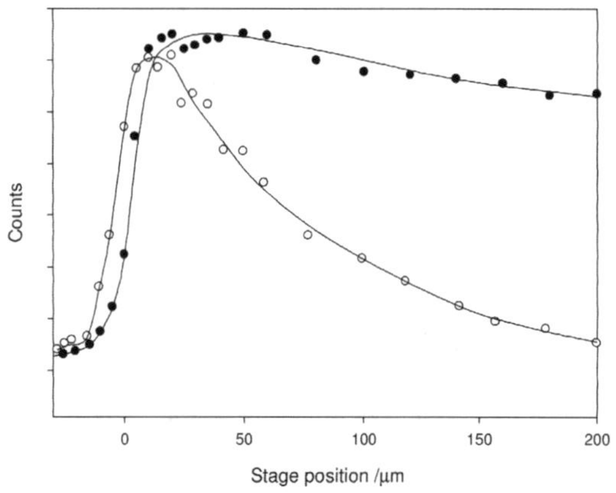

Figure 1. Raman depth profiles of block samples of PE (○) and PS (●)

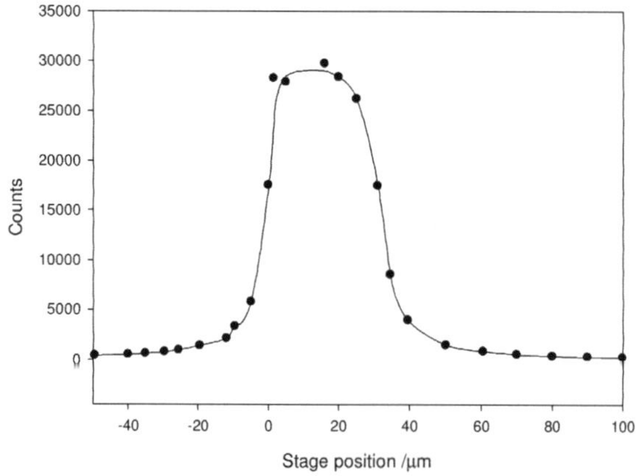

Figure 2. Raman depth profiles of a thin film sample of PS

A void in Sample 1 was probed and analysed using the same optical sectioning technique mentioned above. Spectra shown in figure 3 were gathered at various stage positions ranging from the focal point being at the sample surface to it being deep inside the void. From figure 3 we can see that, at the surface of the sample and for ~100 μm below the surface, the spectrum for PE is obtained. Below this point there is no PE present in the spectra, indicating the presence of a void. Beneath the surface of the sample there is a rise in the spectra at the lower wave numbers. This rise is associated with fluorescence effects and, although chemical data cannot be extracted from this it, indicates ageing [7]. As predicted in figure 1, the signal for PE is slightly greater beneath the surface at a stage height of -65 μm than at the sample surface. It is worth pointing out that there is an additional feature present in all of the spectra at about 550cm^{-1} , which is not present in scans of pure PE but seems to be present throughout the sample including inside the void. Only by further chemical analysis will the exact nature and origin of this feature be determined.

Figure 4 shows the spectra gained by focusing on the surface of Sample 2 at various lateral positions (indicated on figure 5) of a failed void instead of depth profiling a complete void. From this data we can see that region A gives a typical PE spectrum as the laser is focused in a region of undamaged PE. Region B is in the centre of the burst void where there was no sample and so the spectra gained from this region shows nothing other than vestigial PE, peaks associated with material remote from the focal point. In the spectra at low wave numbers, gained from the edge of the void (region C) a PE signal is recognisable but with an increase in background signal as with the spectra in figure 3. Again we take this to indicate fluorescence and hence ageing. Finally in region D, which corresponds to the discoloured region at the edge of the burst void, a large amount of fluorescence and the D and G bands of carbon are present indicating the presence of sp^2 bonded carbon [3]. The signature for region D is characteristic of electrical trees grown in PE. (Note all spectra have been offset from each other by equal amounts along the y axis in order to improve clarity).

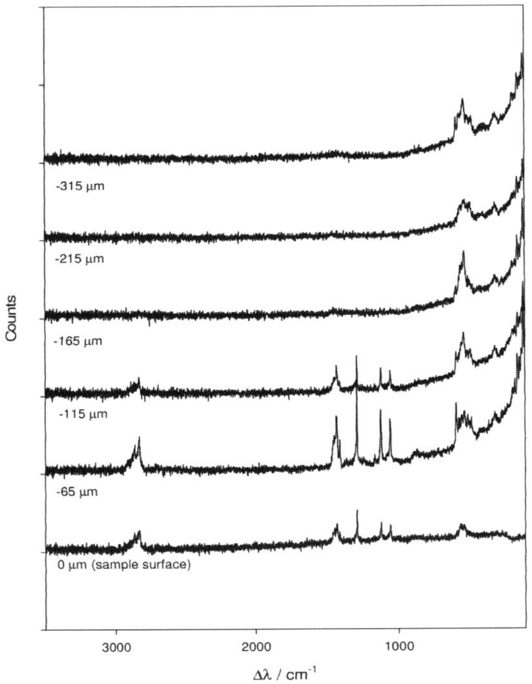

Figure 3. Raman spectra at various stage positions for void in sample 1

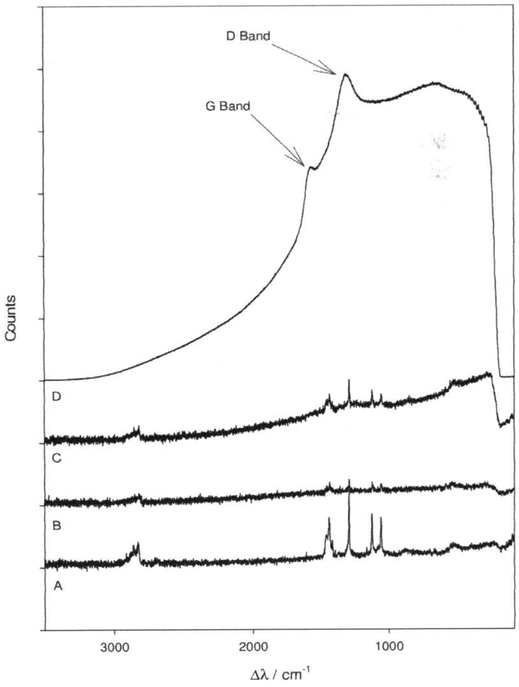

Figure 4. Raman spectra for various lateral positions on failed void in sample 2

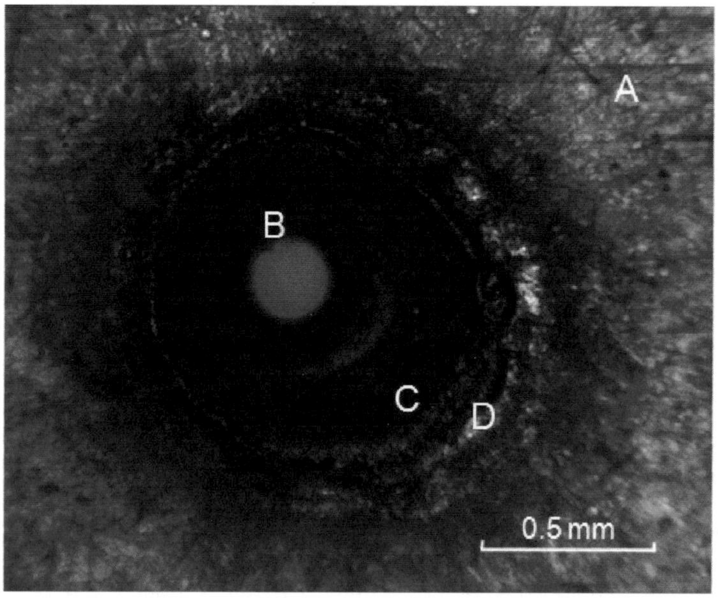

Figure 5. Optical micrograph showing lateral positions analysed in failed void.

4. Discussion

CRMS is an adaptation of established confocal imageing theory which claims that all light from out of focus planes from a transparent/ translucent material is excluded from the final image by using a small aperture or 'pin hole' at the back focal plane of the objective lens. This means that when focused above a sample no data should be attainable and, if the sample is uniform, the data gathered from below the surface should also be uniform. In contrast to this our data shows that it is possible to get a Raman signal when focused above the sample and, with increasing focal depth, the signal attenuates at a rate associated with the clarity of the sample. The results show that detected Raman photons do not come from the focal point alone and that the detected signal is influenced by the optical properties of the sample in question, thus making the exact spatial resolution difficult to determine. Despite these limitations with respect to spatial resolution, depth profiling nevertheless can provide data that demonstrated variations in chemical composition with position in a subsurface void. Potential approaches for improving the spatial resolution CRMS could include the application of beam scattering theory to in confocal geometries which would enable a more complete understanding of the effects of refraction and diffraction on the system. Once the spatial resolution of CRMS can be properly determined, the full potential of its application in the study of subsurface featured in dielectrics can be realized.

Finally, previous Raman studies of electrical trees in PE have yielded the 3 following signatures [2, 3, 8]; pure PE in the regions in-between tree branches, fluorescence in the regions close to the tree channels and the D and G bands of disordered sp^2 carbon in the centre of mature, conducting tree channels. As can be seen from the results presented above, the same three spectral signatures as those found in electrical trees were obtained by a lateral examination of voided samples that have experienced partial discharge activity.

5. Conclusions

In brief, samples of LDPE containing sub surface voids have been examined using CRMS using a depth profiling technique. These results were compared to depth profiles of block and film samples of PE and PS and it was found that the concept of confocal microscopy is not as simple as generally thought and is subjected to inaccuracies in spatial resolution as the Raman photons which contribute to the detected signal emanate from both above and below the focal surface. CRMS has provided data which demonstrates variations in chemical composition with position within voids and revealed that, within a void which has experienced a partial discharge activity, the same signatures as those found in electrical trees are seen. Therefore, despite its limitations, CRMS is a useful tool in the determination of chemical variations beneath the surface of a sample and would be an effective tool in the analysis of ageing and breakdown in dielectric materials.

6. References

1. L. A. Dissado, J.C.F., *Electrical Degradation and Breakdown in Polymers*, ed. G.C. Stevens. 1992, London, UK: Peter Perigrinus.
2. Vaughan, A.S., et al., *On the structure and chemistry of electrical trees in polyethylene.* Journal of Physics D-Applied Physics, 2006. **39**(5): p. 962-978.
3. Vaughan, A.S., S.J. Dodd, and S.J. Sutton, *A Raman microprobe study of electrical treeing in polyethylene.* Journal of Materials Science, 2004. **39**(1): p. 181-191.
4. Tabaksblat, R., R.J. Meier, and B.J. Kip, *CONFOCAL RAMAN MICROSPECTROSCOPY - THEORY AND APPLICATION TO THIN POLYMER SAMPLES.* Applied Spectroscopy, 1992. **46**(1): p. 60-68.
5. Everall, N.J., *Modeling and measuring the effect of refraction on the depth resolution of confocal Raman microscopy.* Applied Spectroscopy, 2000. **54**(6): p. 773-782.
6. Everall, N.J., *Confocal Raman microscopy: Why the depth resolution and spatial accuracy can be much worse than you think.* Applied Spectroscopy, 2000. **54**(10): p. 1515-1520.
7. Sayers, P.W., et al. *Investigation of the structural changes in LDPE and XLPE induced by high electrical stress.* in *Eighth International Conference on Dielectric Materials, Measurements and Applications*. 2000. Edinburgh, UK: Iee.
8. Liu, X.S., A.S. Vaughan, and G. Chen, *A Raman spectroscopic study of bulk and surface ageing phenomena in polyethylene.* 2003 Annual Report Conference on Electrical Insulation and Dielectric Phenomena, 2003: p. 145-148.

Acknowledgements
The authors would like to thank Miss T Brain for her help and assistance in the preparation of the voided samples.

Electrical properties of polyamide/ polyaniline composite films

D Mezdour[1,2,3], M Tabellout[2] , C Vanga Bouanga[2] and S Sahli[1]
[1]Laboratoire de Microsystèmes et Instrumentation, Département d'électronique, Faculté des Sciences de l'ingénieur, Université Mentouri , Route d'Ain el Bey 25000, Constantine, Algérie
[2]Laboratoire de Physique de l'Etat Condensé, UMR CNRS 6087, Université du Maine, Avenue Olivier Messiaen, 72085 Le Mans cedex, France

Email : Dmezdour@Univ-Lemans.fr

Abstract. Surface conductive composites consisting of polyamide 6 (PA-6) polymer films and conductive polyaniline (PANI) particles were chemically elaborated by in situ polymerization of aniline inside the PA-6 matrix, at room temperature. The obtained composites were investigated regarding to their electrical properties. We have measured the frequency dependence of permittivity of these antistatic polymer films (ASPF). Surface conductivity was also characterized using potential decay technique. The enhanced surface conductivity was due to percolation of PANI phase in the surface layer of PA-6 films.
Key words. Conductive composites, Polymers, PANI, Electrical conductivity

1. Introduction

Conducting polymers can be intrinsically conducting or obtained by incorporating conductive particles [1]. However, the non fusibility and non solubility in any solvents of most of them restricted their practical application (except for only few conducting polymers such as poly (3-alkylthiophene), poly (2,5-dialkoxy p-phenylenevinylene) and polyaniline [2], which have found their application in the organic battery field, anti corrosion products (PANI), antistatic protection of photographic films (Polythiophènes) and packages for electronic components protection (PANI) [3]).

Improvements of processability by various attempts to incorporate such conducting polymers in processable insulating matrices have led to remarkable change in electrical conductivity at some concentration [2]. The obtained conductive composites are mainly used for electromagnetic, radio-frequency interference (EMI/RFI) attenuation and in applications that require electrostatic dissipating parts, low power resistance heaters and artificial dielectrics [4].

The insulating nature of Polyamides, which are widely used in microelectronics and aerospace applications may cause significant accumulation of electrostatic charges on their surface, resulting in local heating and premature degradation of electronic components or space structures [5]. This work describes a polyamide/ PANI composite with a level of electrical conductivity superior to that of the polymer matrix sufficient to permit electrostatic discharge. In this paper we discuss the relationship between electrical properties change and filler concentration.

[3] To whom any correspondence should be addressed.

2. Experimental procedure

2. 1. Materials
Polyaniline as a conducting filler and polyamide as the polymer matrix (PM) have been chosen to be the main system for studying the formation process and properties of conducting polymer materials. PA-6 films 25 μm thick are used. Aniline ($C_6H_5NH_2$) was used after additional purification. The oxidant solution is consisting of Hydrochloric acid (HCl) and APS ($(NH_4)_2S_2O_8$).

2. 2. Processing-Formation of the layered conductive composite
The previously weighed film samples were placed in the aniline solution to be swelled during definite period time. The swelling process was performed until the desired aniline film content is reached. Swelled polymer films were then subjected to the chemical aniline polymerization in oxidant solution using a home-built cell (Fig. 1), which permit blocking one side of the PM from the oxidant solution. After the completion of the process, the green transparent film, was washed by distilled water and placed in soxlet apparatus for 24 h to extract with n-hexane (C_6H_{14}) by-products and unreacted aniline. This procedure was followed by drying the film under dynamic vacuum for 24 h up to a stable weight.

2. 3. Characterization
The low frequency (10^{-2}-10^{7} Hz) AC conductivity measurements were made using a Novocontrol broad band dielectric spectrometer. Gold circular electrodes are positioned onto both flat sides of the samples. Electrical conductivity of films surfaces was investigated using corona discharge equipment (Fig. 2) and potential decay technique. The method consists of charging films surface to a desired potential (controlled by the grid potential) and recording potential decay using a programmable 6512 Keithley electrometer [6].

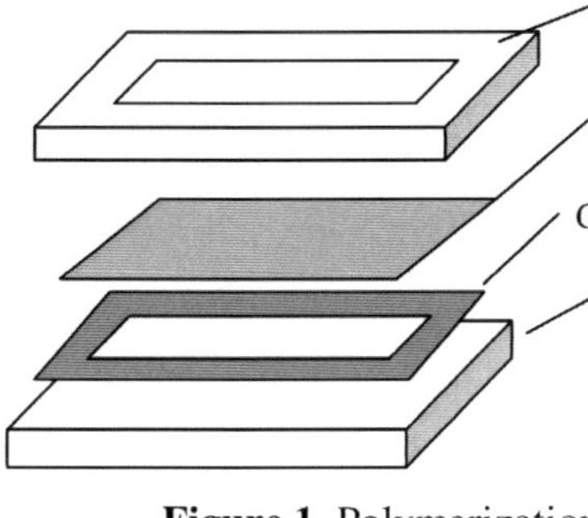

Figure 1. Polymerization cell.

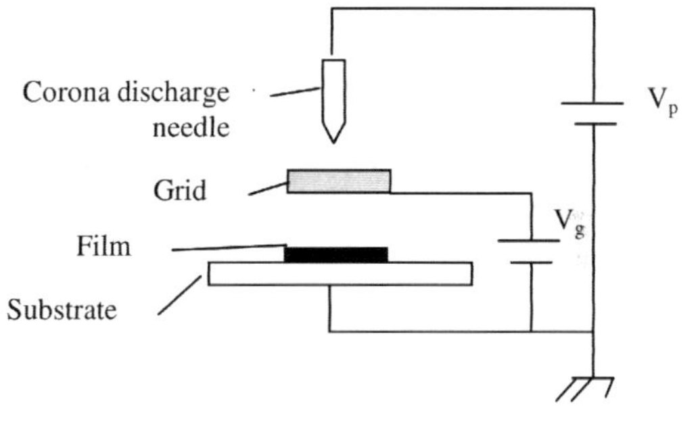

Figure 2. Corona discharge setup.

3. Results and discussion

3. 1. Electrical properties of PA-6/PANI films
Electrical properties of PA-6/PANI films were investigated using the dielectric spectrometer and potential decay technique, the films were sandwiched between two electrodes making the contact with insulating side of the films.

Figure 3 shows the evolution versus frequency of the real part of permittivity ε'of films with different concentrations of aniline at T=-110 °C. When conducting fillers are dispersed in an insulating polymer, the dielectric constant of the composites increases with the volume fraction of fillers as expected by the theory of percolation (Fig. 3- a). The abrupt increase of the dielectric constant is noticed when the volume fraction of the fillers is above 10 wt. % (Fig. 3- b). At each frequency, ε' increases gradually with p, but the changing rate of ε' at a given frequency is not uniform. There is a non linear relation between ε' values and filler volume fraction. This non linear increase of ε' is caused by multiples interactions among particles which become important when particles are closer. Indeed, the dielectric constant of the composites is a function of its capacitance, which is proportional to the quantity of charge stored on either surface of the sample under an applied electric field [7]. In place of pure PA-6, with the composites filled by the conducting fillers, the quantity of the accumulated charge will increase because of the polarization of the PA-6/filler interfaces.

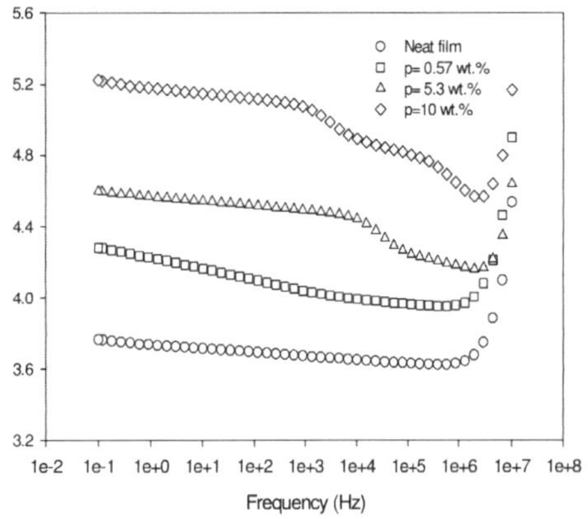

(a)

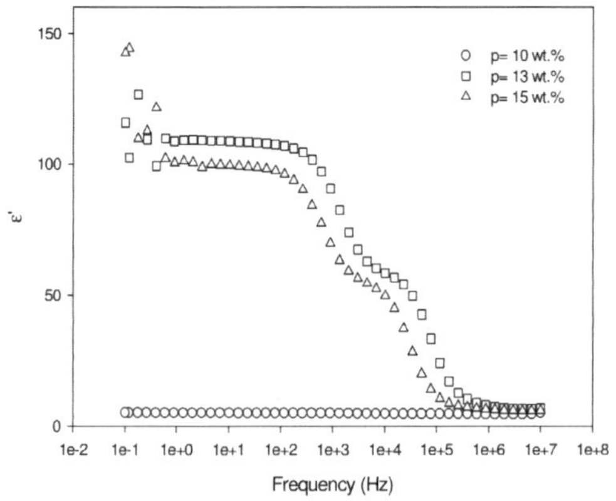

(b)

Figure 3. Evolution of relative permittivity versus frequency of PA-6 films containing different concentrations of aniline at T=-110°C.

The polarization makes an additional contribution to the charge quantity. From this point of view, the dielectric constant of the composites will be higher than that of the pure PA-6 and it will also increase with increasing the concentration of fillers.

For concentrations above 10%, high values of ε' (at low frequencies) were due to sample heterogeneity. This increases interfacial polarization which is noticed by a sharp decrease of ε' at higher frequencies. For these higher concentrations, curves also exhibit relaxation phenomenon, clearly noticed on imaginary part of permittivity ε" curves.

Figure 4 represents the frequency dependence of ε" for composites PA/PANI films. Compared to the virgin film, curves indicate that PANI presence in the film causes a great increase of ε" in the whole studied frequency region and for all PANI concentrations.

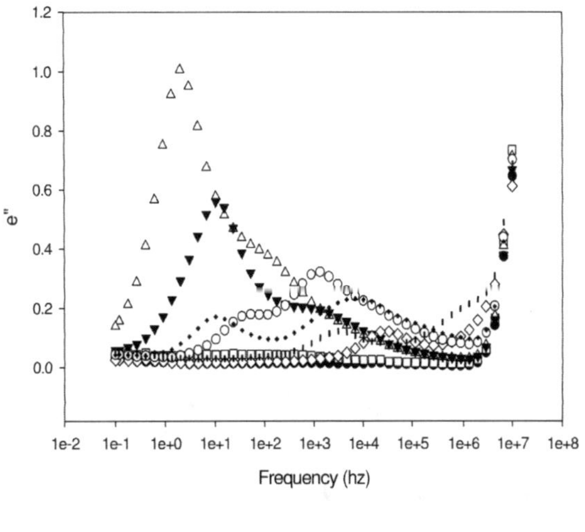

Figure 4. Evolution of ε" versus frequency of PA-6 films containing different concentrations of aniline at T= –110°C, ●: neat film, □: 0.57 wt.%, △: 3 wt.%,▼: 4 wt.%, ○:5.3%, ◇: 8wt.%, + : 9 wt.%, |: 10 wt.%.

The relaxation peaks are attributed to the interaction between PA-6 and doped PANI [8]. Previous work [9] has demonstrated the existence of multiple relaxation peaks due the interface between the composite layer and the virgin pa6 film and to the interface between the pani particles and the matrix. Another relaxation can occur referring to internal phenomena in the PANI particles. The first peak centred at a characteristic frequency f_{max} decreases and shifts toward higher frequency for increasing concentrations of aniline up to 8 wt.%, while the additional relaxation pick revealed at high frequency increases with increasing aniline content and is obviously linked to the presence of charge carriers in the PANI containing layer [10]. Indeed, in conducting polymers and their blends, relaxation processes are assigned to hopping and/or oscillations of charges around fixed pinning centres [11-12]. The increase in the characteristic frequency means shorter relaxation time ($\tau = 1/f_{max}$). This is compatible with an increase in the mobility of charge carriers [12]. This assumption is checked by conductivity measurements. Indeed, the surface conductivity of the films was verified by the potential decay after corona charging of sample's surfaces under 6 kV.

Figure 5 shows evolution of surface potential versus time for PA-6/PANI films. For the film with 1% of aniline concentration one can notice the classical potential decay curve (Fig. 5- a). There are some residual charges on the surface indicating that the film presents some surface electrical resistance to carrier motion. The absence of continuous conductive paths leads to this low electrical conductivity and thus a low relative permittivity as confirmed by dielectric spectroscopy (Fig. 3-a). Increasing concentration of aniline to 4 wt. % enhances the decay of charges which became faster compared to the previous sample. Charges were dissipated in a short time meaning that the surface conductivity was enhanced further. Therefore, a relatively high PANI content is needed to form a PANI network that generates continuous conducting composite. The film of 9 wt. % concentration exhibits a zero surface potential (Fig. 5- b). This can be explained by the dissipation of charges in a very short time due to the enhanced conductivity of the surface proportional to the increased concentration of aniline and thus to that of polyaniline. The lapse of time between the end of charging and the start of measuring was estimated to be about 30 s being sufficient to dissipate all charges. This signifies that surface conductivity will increase considerably for high concentrations and so will do the permittivity versus the frequency for sample containing 10 wt. % of aniline and more due to the percolative pani layer on film's surface. This is also confirmed by dielectric permittivity measurements (Fig. 3-b). These results suggest that these films can be used successfully as antistatic films with a low concentration of aniline.

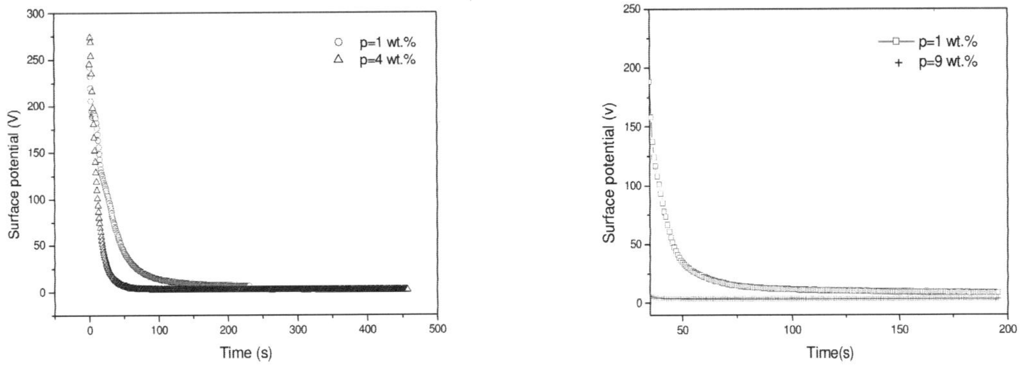

(a) (b)

Figure 5. Potential decay curves on composite film surfaces for various concentration of aniline content (a) Vp=5kV, (b) Vp=6 kV.

4. Conclusions

Surface conductive composites were prepared by *in situ* polymerization of aniline in a polyamide matrix. The later was previously swelled in aniline until reaching a desired concentration. Polymerization was achieved by exposing the film's surface to the oxidant solution. The obtained PA/ PANI composites were studied with respect to their electrical properties using surface potential decay and AC measurement techniques. The composite have the features of a low PANI content and a relatively high surface conductivity. Based on the experimental results, the following conclusions can be made:

(1) The drastic change in the electrical conductivity above some threshold concentration was observed on potential decay curves. These results can be explained in terms of the percolation process.

(2) The electrical conductivity of polyamide/PANI composites increases continuously with increasing PANI content. The conductivity in these composites resulting from the formation of conducting paths in the polymer by the PANI. Depending on the concentration of PANI, it is possible to tailor the conductivity of the composite over many orders of magnitude according to desired utilization technology.

(3) The increase of dielectric permittivity with increasing PANI content is very interesting since large dielectric constant is very important for development of new generation dynamic random access memories and micro-electromechanical systems

(4) The non linear increase of permittivity of polyamide/PANI composites with increasing PANI content is linked to multiples interactions among particles.

(5) Relaxation phenomena noticed on permittivity curves of the composite films at low temperature are attributed to different interfacial polarizations existing in this kind of composites.

5. References

[1] El-Tantawy F, Kamada K and Ohnabe H 2002 *Mater. Lett.* **56** 112-126
[2] Yin X H, Kobayashi K, Yoshino K, Yamamoto H, Watanuki T and Isa I 1995 *Synth. Met.* **69** 367-8
[3] Garnier F 1998 *Electronic Materials: The Oligomer Approach,* ed. K Müllen and G Wegner (Weinheim: Wiley-VCH)
[4] Gokturk H S, Fiske T J and Kalyon D M 1993 *J. Appl. Polym. Sci.* **50** 1891-1901
[5] Ounaies Z, Park C, Wise K E, Siochi E J and Harrison J S 2003 *Comp. Sci.Technol.* **63** 1637-46
[6] Sahli S, Bellel A, Ziari Z, Kahlouche A and Segui Y, 2003 *J. Electrostatics* **57** 169-181
[7] Dang Z-M, Zhang Y-H and Tjong S-C 2004 *Synth. Met.* **146** 79–84
[8] Tabellout M, Fatyeyeva K, Baillif P-Y, Bardeau J-F and Pud A A 2005, *J. Non-Cryst. Solids* **351** 2835–41
[9] Adohi B J-P, Vanga Bouanga C, Fatyeyeva K and Tabellout M 2009 *J. Phys. D : Appl. Phys.* **42**(1) 015302
[10] Pud A A, Tabellout M, Kassiba A, Korzhenko A A, Rogalsky S P, Shapoval G S, Houzé F, Schneegans O and Emery J R 2001 *J. Mater. Sci.* **36**(14) 3355-63
[11] Pinto N J, Sinha G P and Aliev F M, *Synth. Met.* **94** 199–203
[12] Fattoum A, Gmati F, Bohli N, Arous M and Belhadj Mohamed A 2008 *J. Phys. D: Appl. Phys* **41** 095407

Calculation of photonic band-gap of one dimensional photonic crystal

J Zhao[1,2], X Li[1], L Zhong[1] and G Chen[2]
[1]State Key Lab of Electrical Insulation & Power Equipment
Xi'an Jiaotong University, Xi'an 710049 CHINA
[2]School of Electronics and Computer Science
University of Southampton, Southampton SO17 1BJ UK

E-mail: jz208r@ecs.soton.ac.uk

Abstract. Photonic crystal is a composite dielectric with periodic modulation of refractive index that brings about unique photonic band-gap effect by which light could be totally reflected on the surface of such material regardless of any incident angle or mode. Many applications based on this mechanism have been developed and introduced into new waveguides. Such band-gap effect could be understood by investigating the optical performance of one dimensional photonic crystal. In this paper, the principle of band-gap effect of one dimension photonic crystal is firstly explained by Bragg reflection, the band-gap centre is obtained; then the modelling of band-gap of one dimension photonic crystal is conducted as well, the result is in agreement with the Bragg results and shows that the centre of photonic band-gap shifts to larger wavelength with higher refractive index (RI) ratio or dimension ratio between the two component dielectrics, but the involved boundary condition in simulation leads to higher onset RI ratio (3.5) responsible for the presence of photonic band-gap. This work would be useful to the design and fabrication of photonic crystal devices.

1. Introduction

The photonic crystal consists of two dielectrics with different permittivity assumed to be constant [1]. These two dielectrics are arranged alternately in three dimensions. While the space between the two components is reduced to be as small as ~1μm, or even smaller, light beam with a particular frequency (or wavelength) that radiates on the crystal could be totally reflected. The energy responding to such frequency is right located in a forbidden region of its energy band and the light is forbidden transmitting. Such forbidden band is named the photonic band-gap. Materials with photonic band-gap attract increasingly both fundamental and technological interest around the world. The introduction of photonic crystal structure into optical fibres to create superb lower-loss optical fibres and nonlinear optical devices is one classic application [2], which is bringing growing attention.

Understanding of such photonic band-gap effect is of great importance. To that point, the basic one dimensional photonic crystal is chosen to be investigated for simplicity in this paper. The mechanism of photonic band-gap is illustrated by Bragg reflection and then the photonic band-gap is calculated through numerical modelling by the finite difference time domain (FDTD) method.

2. Bragg reflection in photonic crystal

A schematic light flow into one dimensional photonic crystal is presented in figure 1. A and B indicate two dielectrics constructing one dimensional photonic crystal, and their thickness and refractive indices are *a*, *b* and n_1, n_2 respectively. Now the situation of perpendicular propagation of the incident light L_i is considered particularly.

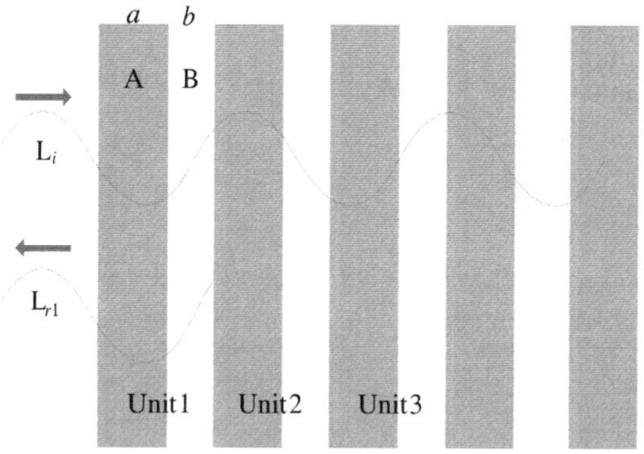

Figure 1. A light flow into one dimensional photonic crystal.

The two dielectrics are assumed to be loss-less within the operational frequency range, so the light beam reaching the internal interfaces undergoes only reflection and transmission without any absorption and scattering. As light energy goes off with each reflection, the eventual light energy that goes out the whole crystal will attenuate to be zero after infinite layers. That means that light could be totally reflected after infinite periodic units. Actually the total reflection can be realised within finite layers if the frequency of light is right in certain band-gap.

As for the photonic crystal in figure 1, the light beam should be reflected at each interface, however, the reflection at the interface of every periodic unit is considered to be dominantly responsible for its unique properties. When the incident light L_i with wavelength λ propagates into one dimensional photonic crystal, it will be reflected at the interface between the first two units and there comes the reflect light L_{r1}; likewise, Light L_{ri} occurs at the interface between following units.

The difference of optical path between the incident light L_i and the reflect light L_{r1} is expressed as,

$$d = 2(n_1 \times a + n_2 \times b) \qquad (1)$$

If the optical path difference *d* satisfies the condition in equation (2), L_i will superpose with L_{r1} and produce a standing wave, the other reflect lights L_{ri} will also join in the standing wave. Then the incident light L_i with wavelength λ is totally reflected, namely the Bragg reflection, and forbidden transmitting through the crystal. The corresponding frequency $f=c/\lambda$ (*c* is the velocity of light in vacuum) is right in an energy region which is forbidden by the photonic crystal and thus the incident light could be totally reflected. Such wavelength λ (or the frequency *f*) can be named as the centre of the band-gap.

$$d = 2(n_1 \times a + n_2 \times b) = m\lambda \quad m = 1,2,3... \qquad (2)$$

The central wavelength λ_c of the band-gap is expressed in equation (3) and it is decided by the refractive index and dimension parameters. When considering the dimensionless parameters, n_1/n_2, a/b, and the crystal lattice constant of the photonic crystal is $l=a+b$, the central wavelength can be expressed as follows,

$$\lambda_c = \frac{2(n_1 \times a + n_2 \times b)}{m} = 2n_2 \frac{l}{1+\frac{a}{b}} \left(\frac{n_1}{n_2} \times \frac{a}{b} + 1\right) \frac{1}{m} \quad m = 1,2.3... \qquad (3)$$

2.1 Influence of refractive index ratio on central wavelength

Given that one dimensional photonic crystal is composed of dielectric A and air ($n_2 \approx 1.0$), the crystal lattice constant is l=10μm, the dependence of central wavelength on the refractive index ratio $n_1/n_2=n_1$ is shown in figure 2 (m=2).

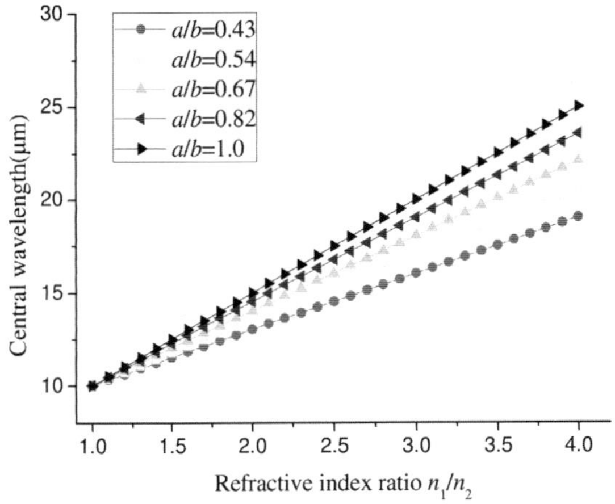

Figure 2. Dependence of central wavelength on refractive index ratio.

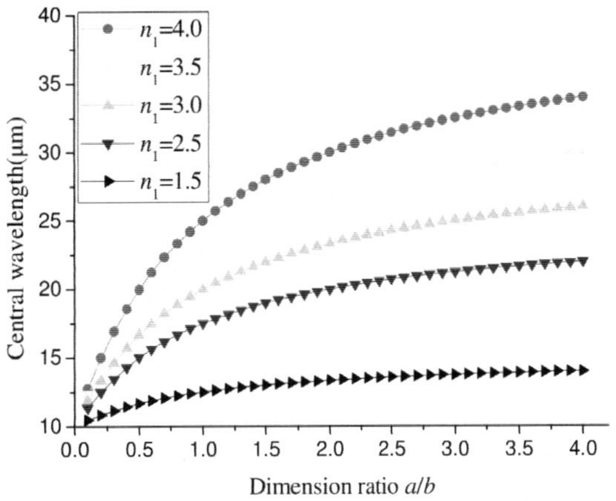

Figure 3. Dependence of central wavelength on dimension ratio.

As indicated in figure 2, the central wavelength increases with the rise of refractive index ratio n_1/n_2, even when the ratio reaches 4.0, that is possibly the highest RI of silicon [3]. This suggests that photonic crystals can be made of materials ranging from transparent polymers to semiconductors.

2.2 Influence of dimension ratio on central wavelength

Given that the crystal lattice constant l=10μm, i.e. $a+b$=10μm; the refractive index of dielectric A is n_1=1.5, 2.5, 3.0, 3.5, 4.0, and the other dielectric is still air ($n_2 \approx 1.0$), m=2. The dependence of central wavelength on the dimension ratio a/b is shown in figure 3, which clearly expresses a non-linear increasing with dimension ratio a/b. It indicates that the thinner layer dielectric A with higher index has much influence than B on the central wavelength.

3. Simulated photonic band-gap

To verify the results obtained above, numerical simulation of the band-gap of one dimensional photonic crystal based on the FDTD method is conducted. The layout of one dimensional photonic crystal is shown in figure 4. The photonic crystal consists of dielectric A and B (air), and it is set to the right side of the excitation source in horizontal direction. The thickness of the dielectric A and air is a, b respectively, the lattice constant is l=10μm. a is increased from 0.5 μm by 0.5μm step until a/b=1.0. The refractive index of the dielectric A n_1 varies from 1.0 to 4.0.

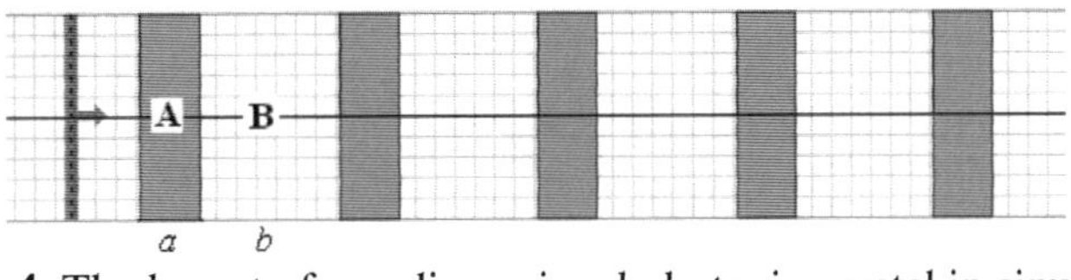

Figure 4. The layout of one dimensional photonic crystal in simulation.

The band-gap properties (indicated by wavelength not frequency) of one dimensional photonic crystal is calculated and shown in figure 5, figure 6. The width of band-gap increases nearly linearly with the refractive index ratio between the dielectric A and air; but the band-gap start to appear when the index ratio is above 3.5, which is different from the prediction by Bragg reflection. In figure 6, the width of band-gap shows a maximum value when the dimension ratio is around 0.6, which apparently does not change at various refractive index ratios. And it goes higher with larger index ratio.

In addition, the centre of band-gap is also calculated. As shown in figure 7 and figure 8, the numeric results indicate same trend of dependence of centre of band-gap on the refractive index ratio and dimension ratio as in figure 2 and figure 3, except that band-gap starts to turn up at n_1=3.5, and the band-gap goes off at lower dimension ratio.

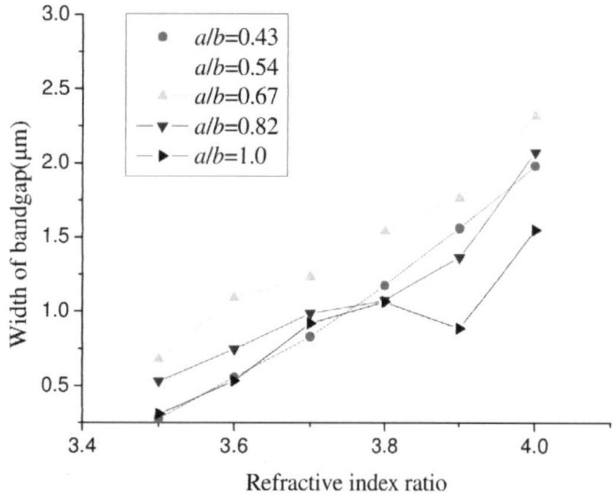

Figure 5. Width of band-gap vs. index ratio.

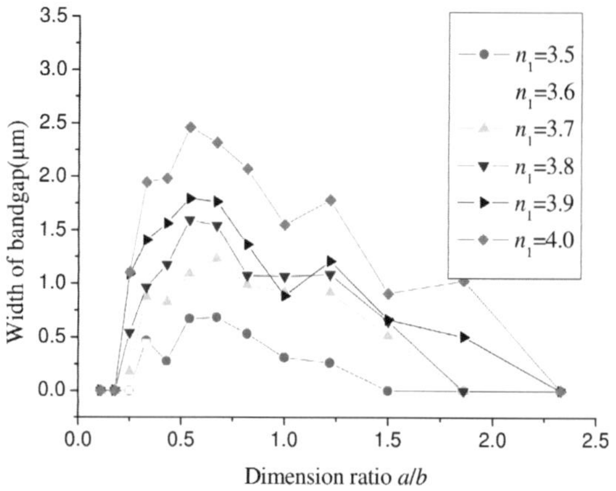

Figure 6. Width of band-gap vs. dimension ratio.

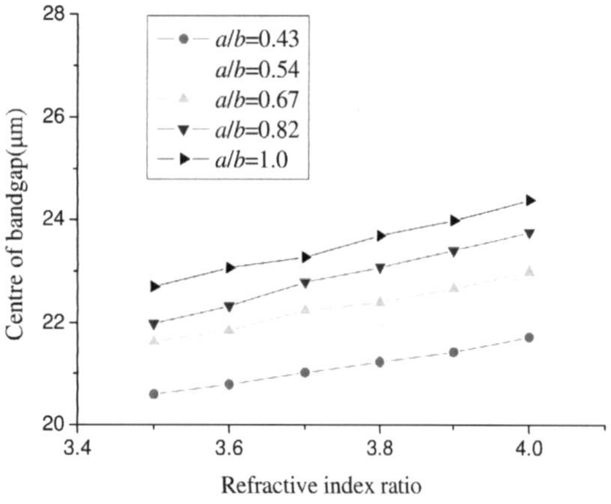

Figure 7. Centre of band-gap vs. index ratio.

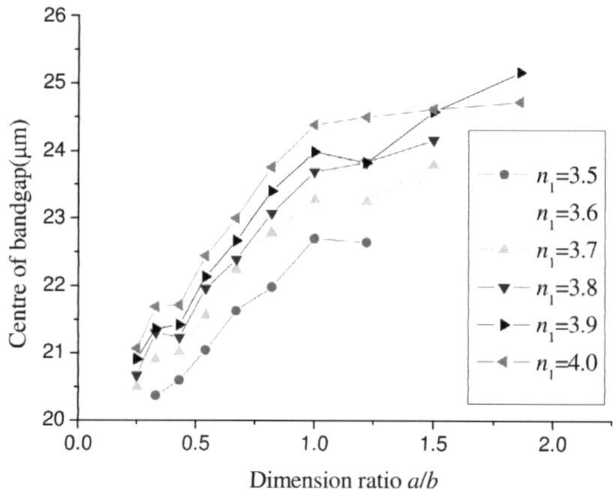

Figure 8. Centre of band-gap vs. dimension ratio.

4. Conclusion

Numeric results validate the analysis on the formation of photonic band-gap through Bragg reflection within one dimensional photonic crystal and verify that the equation of central wavelength can predict the possible band-gap centre even though indicated by wavelength not frequency, as well as the appropriate value of m=2 for the selected model. However the numerical analysis presents rigorous result that higher refractive index is required to obtain band-gap.

5. References

[1] Deych L I, Erementchouk M V et al. 2005 *Phys. Stat.* sol (c) **2** 805-808.
[2] Tang R 2005 "Design and fabrication of photonic crystal fiber." *Dissertation of master, Xi'an Jitao-tong Univerisity.*
[3] Aspens D E 1988 *Properties of Silicon* INSPEC, IEE (*London, UK*) ch **2** p 59-80.

Electrospinning atactic polystyrene: a neutron scattering study

S Mohan, S Sen, G R Mitchell, R H Olley, F J Davis
Polymer Science Centre, University of Reading, Reading RG6 6AF UK

E-mail: s.d.mohan@reading.ac.uk

Abstract. Electrospinning is a method used to produce nanoscale to microscale sized polymer fibres. In this study we electrospin 1:1 blends of deuterated and hydrogenated atactic-Polystyrene from N,N-Dimethylformamide for small angle neutron scattering experiments in order to analyse the chain conformation in the electrospun fibres. Small angle neutron scattering was carried out on randomly orientated fibre mats obtained using applied voltages of 10kV-15kV and needle tip to collector distances of 20cm and 30cm. Fibre diameters varied from 3μm – 20μm. Neutron scattering data from fibre samples were compared with bulk samples of the same polymer blend. The scattering data indicates that there are pores and nanovoiding present in the fibres; this was confirmed by scanning electron microscopy. A model that combines the scattering from the pores and the labelled polymer chains was used to extract values for the radius of gyration. The radius of gyration in the fibres is found to vary little with the applied voltage, but varies with the initial solution concentration and fibre diameter. The values for the radius of gyration in the fibres are broadly equivalent to that of the bulk state.

1. Introduction

Electrospinning is a technique used to produce polymer fibres varying from nanoscale to microscale diameters. A polymer is dissolved in a suitable solvent, preferably one with a high dielectric constant. This solution is then placed into a spinneret where a high voltage is applied to the spinneret tip. The applied field causes the droplet at the tip of the spinneret to form into a Taylor cone [1]; a jet is then pulled from the Taylor Cone to a grounded collector. As the jet travels through the air the solvent evaporates, and the polymer fibre is deposited onto the grounded collector. Many factors affect the fibre diameter, such as the applied voltage, the distance between the collector and spinneret tip, and the solution concentration. In the electrospinning process, chain entanglements play a critical role in the formation of fibres [1]. We have used small angle neutron scattering to obtain values of the dimensions of the chain trajectories in fibres prepared by electrospinning as part of a program to understand the molecular behaviour during the electrospinning process.

Small angle neutron scattering (SANS) is a technique that can be utilised to analyse polymer chains, in particular the polymer-chain trajectory, at lengths scales of between 10 and 100nm. The technique requires isotopic labelling to gain a contrast, but by utilising polymers selectively substituted with deuterium rather than hydrogen the technique remains a powerful analytical tool for the determination of polymer chain dimensions, both in the bulk state and in solution. [2].

In this study we prepare fibres in 1:1 blends of hydrogenated and deuterated atactic-Polystyrene (a-PS) spun from the solvent N,N-Dimethylformamide (DMF), which is known to be an effective solvent for electrospinning polystyrene fibres [3]. The contrast between the deuterated and hydrogenated polymer chains allows for the study of how the electrospinning process affects the chain conformation.

2. Materials and Methods

2.1 Polystyrene

We have used hydrogenated and deuterated atactic-polystyrene with molecular weights shown in Table 1 as measured by GPC.

Table 1. Characteristics of Polymers

Polymer	Label	*Molecular Weight (g/mol)*	Mw/Mn
Hydrogenated Polystyrene	HPS-1	241,000	4.8
Hydrogenated Polystyrene	HPS-2	136,000	4.4
Deuterated Polystyrene	DPS-1	205,000	3.8
Deuterated Polystyrene	DPS-2	128,000	3.8

2.2 Solutions

For electrospinning 1:1 blends of HPS-1/DPS-1 and HPS-2/DPS-2 in DMF at 0.2, 0.3 and 0.4 g/mL of solvent were prepared. High concentrations were used so that smooth fibres were formed without beading [1].

2.3 Electrospinning

The electrospinning equipment is arranged in a horizontal set up. The syringe pump equipped with a needle gauge of 22G (0.41mm inner diameter) provided a flow rate of 0.13ml/min. Voltages of 10 kV, 12.5 kV and 15 kV were applied at distances from tip of the needle to grounded collector of 20 cm and 30 cm. The grounded collector was a sheet of aluminium foil and the fibres were collected as randomly orientated fibre mats. NMR analysis on fibres immediately after spinning showed small levels of solvents. These were eliminated by standing at room temperature for 24 hours after spinning and all samples for neutron scattering measurements were prepared in this manner

2.4 Small Angle Neutron Scattering (SANS)

SANS data is given in the form of the differential scattering cross section ($(d\Sigma/d\Omega)(Q)$ (cm^{-1})), as a function of the scattering vector ($Q(\text{Å}^{-1})$). The differential scattering cross section contains all the information on the size, shape and interaction between the scattering centres within the sample [2]. SANS was performed using LOQ at the STFC ISIS Facility in the UK. A wavelength range of 2.2 Å to 10 Å was employed giving a scattering vector range $|Q|(\text{Å}^{-1})$ range of 0.006 Å^{-1} – 0.27 Å^{-1} where $|Q|$ is given by (1) and 2θ is the scattering angle. The data was collected on a 2 dimensional detector which was radially averaged to give $(d\Sigma/d\Omega)(Q)$ (cm^{-1}) as a function of $|Q|$ (Å^{-1}).

$$|\underline{Q}| = 4\pi \sin\theta / \lambda \qquad (1)$$

3.Results and Discussion

3.1 SANS on Fibres

Scattering from the randomly orientated fibre mats shows a strong component from pores and nanovoids. Thus, scattering from purely deuterated fibres shows that ($(d\Sigma/d\Omega)(Q)$) varied as Q^{-4} which indicates sharp interfaces are present within the material. This is confirmed by analysis of fibres in a scanning electron microscope. The fibre surfaces show varying amounts of porosity and fractured fibres show voiding present inside, with pores varying from 20nm-120nm in diameter. Fibre mats exhibit a distribution of fibre diameters and in order to get a statistical representation for the mean average diameter a grid is overlaid over the image, and fibres are randomly selected in each square of the grid. Figure 3 shows the diameter distribution of fibres obtained in this way for fibres prepared from a 0.2 g/mL solution.

Figure 1. SEM micrograph of a randomly orientated fibre mat prepared from a 0.2g/mL solution

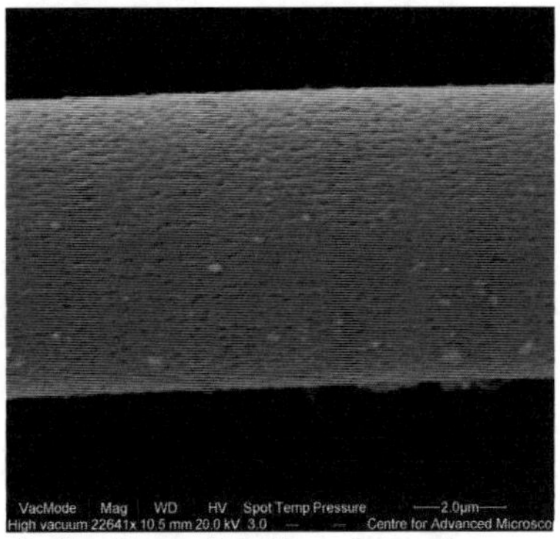

Figure 2. Surface of a fibre from the mat shown in Figure 1 showing porosity. Prepared from a 0.2g/mL solution

The amount of porosity varies between samples. Figure 2 is a case in which a high surface porosity is present on the fibres. The relative humidity level is thought to contribute to porosity in electrospun fibres [4], in this case samples were prepared at humidities of between 47%-50%. Voiding is caused by the high concentration of the polymer and the large fibre diameter trapping solvent into the fibre [5]. The results indicate that with increasing solution concentration there is an increase in fibre diameter.

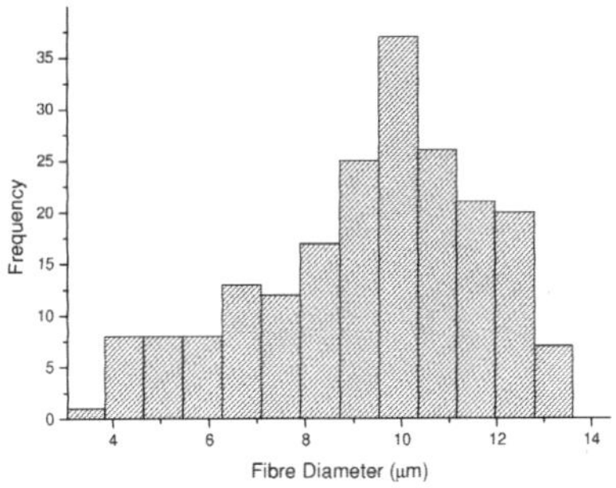

Figure 3. Distribution of fibre diameters measured in a randomly orientated sample spun from a 0.2g/mL solution..

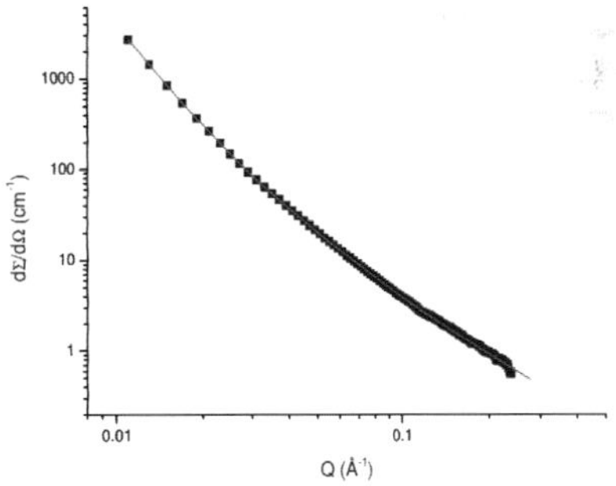

Figure 4. Log scale of scattering from fibre sample (squares) and the fit using equation (2) (line). The fibres were spun from a 0.2g/mL solution.

We have modelled the scattering from fibres using the Debye Gaussian Coil plus the scattering component from the pores and this can be described by Equation (2), where $A = x\,(1\text{-}x)V(\Delta\rho)^2$, x is the concentration of the labelled polymer, V is the volume, $\Delta\rho$ is the difference in the scattering length and B is a scaling factor relating to the fraction of the voids [6].

$$I(Q) = \frac{2A}{(QR_g)^4}\left\{(QR_g)^2 - 1 + e^{-(QR_g)^2}\right\} + BQ^{-N} \quad (2)$$

Using this model with the scattering data we are able to extract values for the radius of gyration (R_g). Optimisation of the model to the data was achieved using the data-fitting program FISH [6]. The radius of gyration of the polymer chains in the fibres was found to decrease with increasing initial solution concentration as shown in Figure 5, and replotted in Figure 6, which shows that there is an increase in the R_g of the polymer chains in the fibre with decreasing fibre diameter. We expect that this is related to the different solution concentrations rather than a direct consequence of the fibre diameter. No variation in R_g was found with other changes to the preparation conditions, such as the applied voltage and the collector distances. The values of the radius of gyration obtained for the fibres are broadly equivalent to those of the bulk state (R_g ~ 90Å) suggesting that the chains have largely reached an equilibrium state.

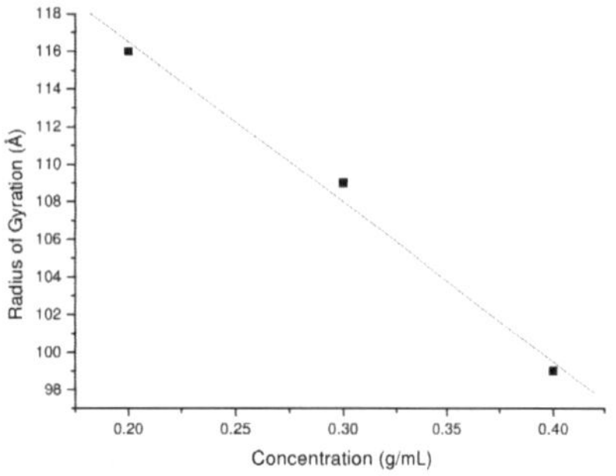

Figure 5. Radius of gyration for fibre samples versus the initial solution concentration. (Solid line acts as a visual guide)

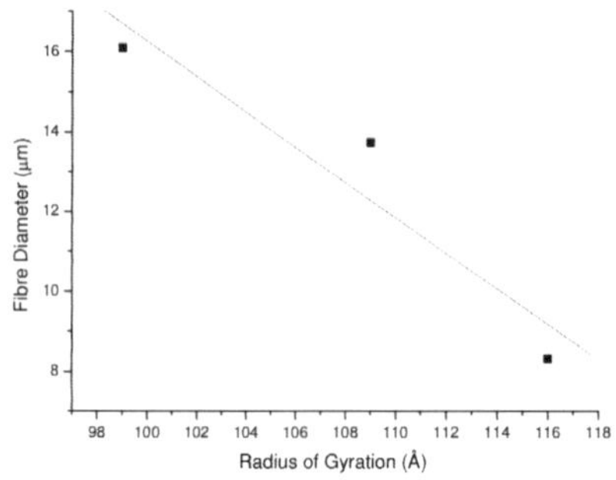

Figure 6. Variation of the fibre diameter with radius of gyration of the polymer chains. (Solid line acts as a visual guide)

4. Conclusions

Electrospun fibres of isotopically labelled polystyrene were successfully studied by SANS and the conformation compared to that of the solution state. The scattering from the fibres indicates high porosity and nanovoids, which was also confirmed with scanning electron microscopy. A model was successfully applied to describe the scattering from the pores and the labelled chains allowing for a value for the radius of gyration to be extracted from the scattering data. It was found that there is little variation in the radius of gyration given the voltages and distance that were utilised in this case. The radius of gyration decreases in the fibre with increasing fibre diameter as the solution concentration increases. The radius of gyration for the fibres approaches that of the bulk state.

5. References:

[1] Ramakrishna S, Fujihara K, Teo WE, Lim TC and Ma Z 2005 *An Introduction to Electrospinning and Nanofibres* World Scientific Publishing Co
[2] Pethrick RA and Dawkins JV 1999 *Modern Techniques for Polymer Characterisation,* John Wiley
[3] Jarusuwannapoom T, Hongrojjanawiwat W, Jitjaicham S, Wannatong L, Nithitanakul M, Pattamaprom C, Koombhongse P, Rangkupan R, and Supaphol P 2005 *European Polymer J.* **41** 409-21
[4] Medeiros ES, Mattoso LHC, Offeman RD, Wood DF and Orts WJ 2008 *Canadian J. Chem.* **86** 590-99
[5] Kim K, Kang M, Chin IJ and Jin HJ 2005 *Macromolecular Research* **13** 533-37
[6] Heenan RK 2005 *RAL Report* 89-129

Conducting nanofibres produced by electrospinning

S.Sen, F.J. Davis,[1] G.R. Mitchell and E. Robinson
Polymer Science Centre, University of Reading, Whiteknights, Reading RG6 6AD

E-mail: f.j.davis@reading.ac.uk

Abstract. Electrospun fibres based on polypyrrole have been prepared. The incorporation of preformed polypyrrole into fibres electrospun from a carrier polymer can only be achieved when materials are prepared with particulates smaller than the cross-section of the fibre; even so there are some problems, with the substantial loss of material from the electrode tip. As an alternative approach, soluble polypyrroles can be prepared but these are not of sufficient viscosity to prepare electrospun fibres, once again a carrier polymer must be employed. More effective loadings are gained by the process of coating the outer surface of a pre-spun fibre; in this way electrospun fibres coated with polypyrrole can be prepared. This approach has been adapted to produce silver coated polymer fibres by the use of copolymers of styrene and 3-vinyl benzaldehyde.

1. Introduction

Electrospinning is a technique which relies on the build-up of electrostatic forces on the surface of a drop to eject polymeric material, which on evaporation of the solvent, produces polymeric fibres with dimensions (of the order of 10s of nanometres to a few microns) largely inaccessible by alternative means [1]. The generation of such fibres has important implications in the field of nanotechnology in general [2], but more specifically in terms of potential applications; these include textiles, filters, catalysts, and a range of biomedical applications [3]. One particularly attractive possibility is the generation of conducting fibres with nanometre dimensions, and a number of workers have explored this possibility, through, for example, the incorporation of polyaniline [4] or through chemical or physical vapour deposition techniques [5]. Recently particular interest has been shown in the incorporation of polypyrrole into electrospun polymers, either through the introduction of the material prior to spinning, in a soluble form [6], [7], or through surface coating polymer composites containing an oxidant [8]. Polypyrrole is a particularly attractive material because of its ease of synthesis, the stability of the polymer and the possibility of chemical modification (for example through the incorporation of liquid crystalline groups [9]).

In this presentation we describe routes to producing conducting electrospun fibres. We describe initially routes towards the incorporation of polypyrrole into the fibres, through the dispersion of particles. Subsequently we explore the incorporation of a soluble polypyrrole derivative into fibres, and finally consider routes to the chemical deposition of polypyrrole on the surface. We further explore surface treatments by the preparation of copolymers of styrene with a reactive comonomer; in this case 3-vinylbezaldehyde. This method gives a more even surface coating for oxidised materials. This is demonstrated by reacting the aldehyde-containing polymer with a solution of silver as a reducing agent. The resulting material when analysed using SEM and in particular EDX is shown to

[1] To whom any correspondence should be addressed.

give an even surface coating, with control of the silver content being governed by preparation conditions.

2. Experimental

2.1 Materials
All materials were obtained from commercial suppliers and used without further purification except 3,4-diethylpyrrole, which was synthesised by a literature procedure [10], [11] and styrene copolymers, which were synthesised by free-radical polymerisation of mixtures of styrene and 3-vinylbenzaldehyde under reduced pressure at 55 C using azo-bis(isobutyronitrile) as the initiator. Pyrrole and 3,4-diethyl pyrrole were purified prior to use by distillation over calcium hydride under reduced pressure. The Polypyrrole synthesized using PVA as the steric stabiliser was isolated after centrifugation and then re-dispersed in 10ml of a 10% PVA solution; the solution was then stirred for 24hrs before electrospinning.

2.2 Electrospinning
Electrospinning was performed using a glass syringe mounted in a syringe pump fitted with a 22 gauge needle of length 2.5 cm and with an internal diameter of 0.71 mm together with a flat aluminium electrode placed normal to the needle at distances varying from 10 to 40 cm. A Glassman High Voltage power supply was used which allowed defined voltages over the range 7.5-20 kV.

2.3 Electron Microscopy
All electron microscopy was performed using a FEI Quanta FEG 600 Scanning Electron Microscope (SEM). Samples were coated with a thin layer of metal (gold) prior to analyses under high vacuum at an accelerating potential of 20kV and a working distance of 10mm. Alternatively, non-conducting samples were occasionally analysed under low-vacuum conditions without coating, when required, at an accelerating potential of 12.5kV and a working distance of 8mm. The SEM was fitted with an Oxford Instruments Energy dispersive X-Ray spectrometer which was used to confirm the presence of certain elements in addition to elemental mapping.

3.Results and Discussion

3.1 Dispersion of pyrrole particles.

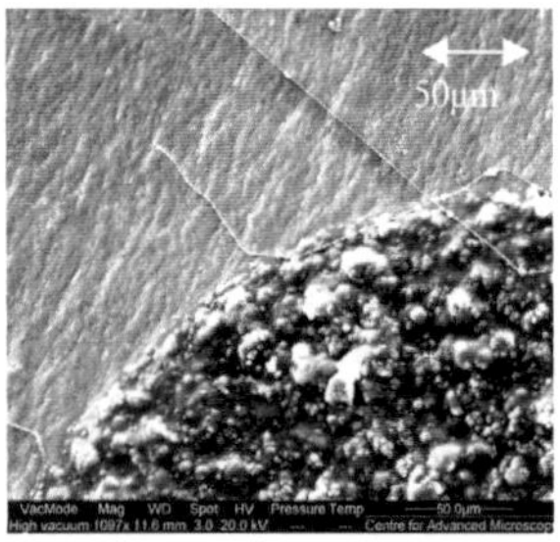

Figure 1: Portion of collector surface after attempts to electrospin chemically prepared polypyrrole

Figure 2: Polypyrrole prepared chemically after grinding in a ball mill.

Although previous workers have suggested that simple linear polypyrrole is not suitable for electrospinning, there is strong evidence that it is possible to produce fibres containing non-soluble materials through electrospinning a dispersion of solid particles in a polymer solution (see for example [12]). To this end we undertook preliminary experiments in which pyrrole prepared chemically through oxidation with either cerium (IV) or iron (III) was dispersed in an aqueous solution of either poly(vinylalcohol) or poly(ethylene oxide). In line with expectations no evidence of fibres was obtained using a range of preparation and electrospinning conditions; instead there was substantial sparking during the process, solvent loss from the needle tip, and spots formed on surface of collector [figure 1] in some cases these then evolved spherulictic structures. On inspection of the chemically prepared polymer using SEM, it was found that the particulates produced were larger than the diameter typically obtained for the carrier polymer fibres (at the appropriate conditions), even after the material had been ground in a ball mill (figure 2).

In view of this problem, materials were prepared with controlled particle size. There are a number of approaches to this, but a particularly attractive one is through the use of poly(vinylalcohol) as a dispersant [13]. Using a modification to the literature method, polypyrrole with a particle size of between 50nm and 120 nm was prepared. In order to increase the concentration of the polypyrrole the material was isolated and re dispersed. In this way poly pyrrole fibres with an average diameter of 300 to 600nm were prepared; EDX analysis was used to map iron and chlorine in the fibres and confirmed the presence of polypyrrole on the basis of the presence of $FeCl_3$ in the polymer as shown in figure 3

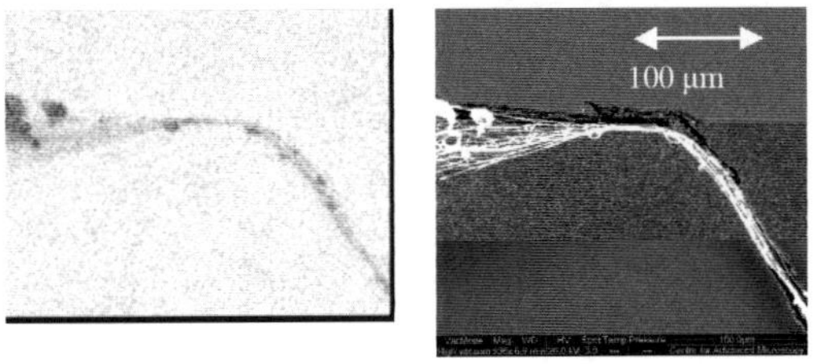

Figure 3. PVA/Polypyrrole composite fibres (1:1, centre) with x-ray mapping showing chlorine content (left) and iron content (right).

3.2 Soluble pyrrole

As an alternative to the addition of solid pyrrole-based particulates, the use of a modified pyrrole with solubilising features was considered. One example of this approach is to use poly(3,4-diethylpyrrole) prepared by oxidising 3,4-diethyl pyrrole with $FeCl_3$, as shown in figure 4. This approach yielded a polymeric material which was substantively soluble in a range of solvents. Homogenous solutions of up to 15%wt/vol in DMF were prepared and then used in the electrospinning process. However, no fibre structures were observed on the collecting electrode for a variety of tip to collector distances and working voltages; presumably the viscosity of the polymer was too low. Mixing with a carrier polymer, in this case polystyrene, in different ratios produced composite fibres with a deep brown/black colouration. SEM in conjunction with x-ray mapping confirmed the presence of Fe in the polymer. In this way it proved possible to produce materials with ratios of the pyrrole derivative of up to 1:2 w/w.

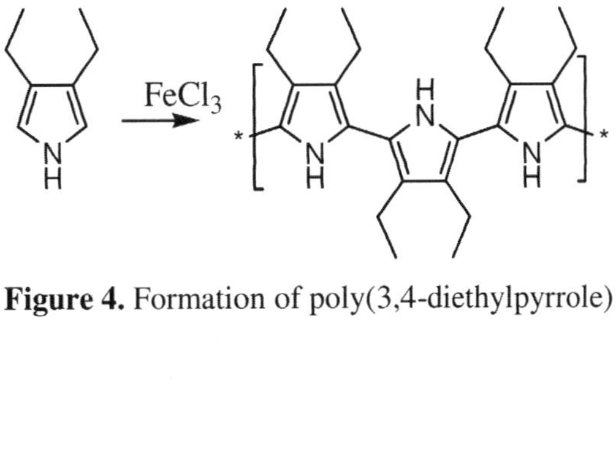

Figure 4. Formation of poly(3,4-diethylpyrrole)

3.3 Surface coating

Although both the approaches described above can be used to introduce polypyrrole into electrospun fibres, in both cases the amount of pyrrole incorporated is rather limited. This has clear consequences if high conductivity is required. As an alternative approach the formation of polypyrrole after the initial formation of the electrospun fibres is an attractive proposition; although weight for weight the loadings of conducting material might be equally small, the concentration of this on the surface will facilitate conductivity. There are a number of approaches to this; the simplest is to place fibres in a solution containing pyrrole and oxidant. Unfortunately this gave a rather poor and uneven coverage. Of a range of systems studied it was found that the most uniformly coated fibres were obtained when polystyrene was dissolved in DMF along with variable amounts of ferric chloride and then electrospun to obtain yellow coloured composite fibres of polystyrene and ferric chloride (figure 5). These fibres were then exposed to pyrrole at 60°C for 4-5 hours as shown in figure 6 (if the exposure to pyrrole vapour was substantially longer then disintegration of the fibre occurred as pyrrole vapour dissolved away the polystyrene).

Figure 5: Polystyrene/$FeCl_3$ fibres electrospun from DMF

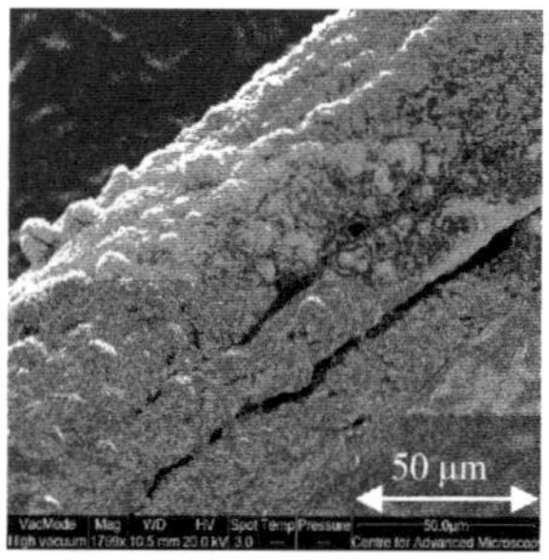

Figure 6: Polystyrene/$FeCl_3$ fibres electrospun from DMF, coated with Polypyrrole;,

The dissolution of the fibres observed during the preparation, as described above, leads to a route towards pure polypyrrole fibres. The composite fibres of polystyrene/polypyrrole were suspended in DMF at room temperature for a period of 24hrs, in order to selectively dissolve the polystyrene leaving behind a hollow polypyrrole fibre. Although in most cases the fibre structure broke down to give black particulates, for selected samples, a hollow fibres was obtained as shown in figure 7. The thickness of the polypyrrole layer shown in this example is approximately 300 to 400nm.

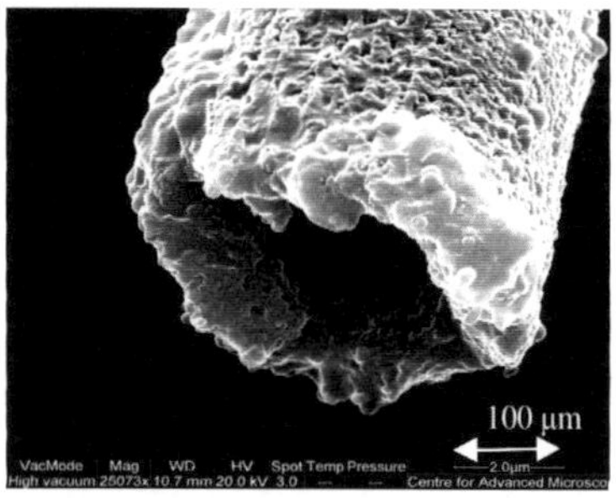

Figure 7: Hollow polypyrrole fibre prepared by suspending a Polystyrene/Polypyrrole composite fibre in DMF at room temperature for 24hrs

3.4 Silver coated fibres pyrrole

An alternative oxidant for pyrrole is silver (I), our attempts to use this oxidant, by for example incorporating silver nitrate into polystyrene, whilst less promising for the generation of polypyrrole than the ferric chloride approach described above, did suggest a process which might provide an alternative type of conducting fibres; namely silver coated fibres. Thus, the methodology applied above was modified to the preparation of metallic coated fibres utilising the reaction between ammonical silver nitrate and an aldehyde (figure 8), commonly known as Tollen's test [14]. In this case investigations suggested that improved coatings were obtained by incorporation of the oxidant (typically glucose) as a co solute during the electrospinning process.

In an extension to the silver coating approach, a range of copolymers were prepared based on polystyrene and poly(vinylbenzaldehyde) of the type shown in figure 9. These were prepared with compositions ranging from 10-50% aldehyde (w/w). Though there were some problems with partial cross-linking of the material containing greater proportions of the reactive unit, these materials were all spun successfully to form fibres (with typical diameters of *ca.* 2 μm). The quality of the silver coating was gauged by SEM and EDX. This was found to be highly dependant on the preparation conditions. However, for the conditions tried (concentration of reagent temperature etc.) it was noted that the most even coated material was one where the fibre contained *ca.* 50% aldehyde (Figure 10), while that with only 10% showed a rather more uneven distribution (figure 11). In this case attempts to remove the central core of the fire (*cf.* figure 7) were unsuccessful. We attribute this to a rather thin coating, consistent with the oxidant being covalently bound to the polymer surface, with little possibility to diffuse through the silver coating.

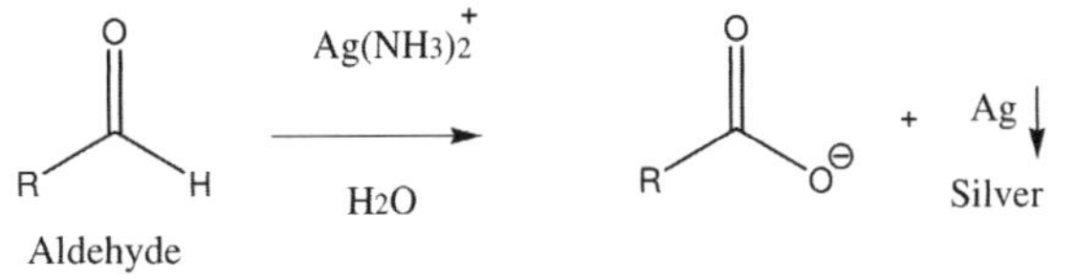

Figure 8: Reaction between Silver(I) and an aldehyde

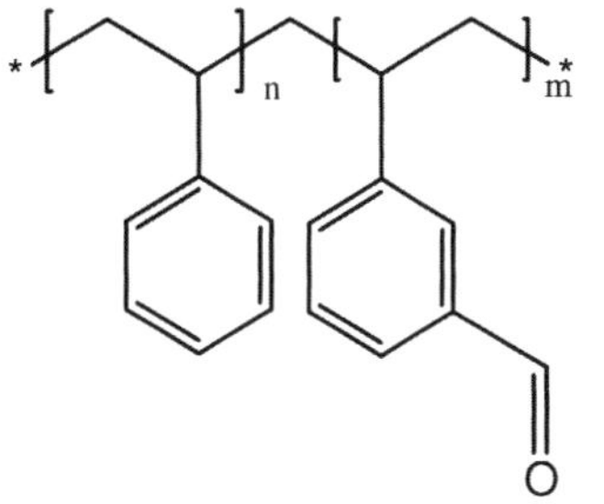

Figure 9:Styrene/3-vinyl benzaldehyde copolymer

Figure 10: silver coated polystyrene fibre from a 50:50 styrene vinylbenzaldehyde copolymer

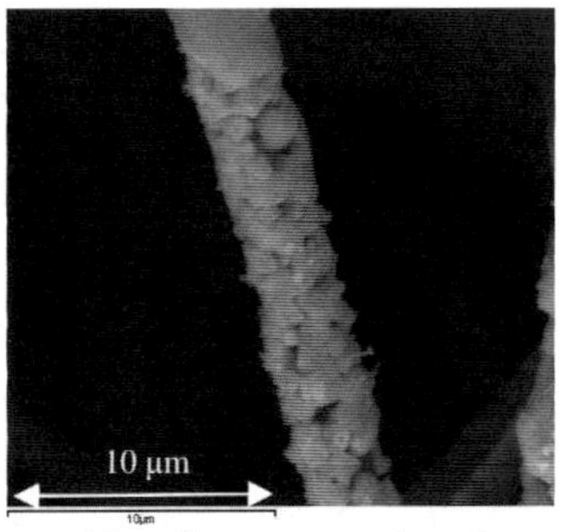

Figure 11: silver coated polystyrene fibre from a 50:50 styrene vinylbenzaldehyde copolymer

4 Conclusions

A number of routes to composite electrospun fibres based on polypyrrole have been described. Polypyrrole particles can be incorporated provided that the dimensions of the particle are commensurate with the fibre diameter; large polypyrrole particles prevent the formation of fibres. However loading with sufficient concentrations of polypyrrole to render the fibres conductive may have consequences on the electrospinning. Soluble pyrrole-based polymers can also be incorporated into similar fibres, however the loadings may be limited and the low viscosity of the pure polymer solutions, at least in these investigations, prevented direct spinning of pure polypyrrole solutions. Surface coating offers a particularly versatile option, the results being particularly promising when the oxidant is contained in the fibres. In some cases part of the central core could be removed to produce essentially a pure polypyrrole hollow fibre. This approach has been extended to the production of oxidising copolymers; such polymers produce particularly uniform silver coated fibres. This post-processing approach offers a promising route to electrically conducting electrospun fibres.

5 References

[1] Ellison C J, Phatak A, Giles D W, Macosko C W and Bates F S, 2007, *Polymer* , **48**, 3306-16.
[2] Reneker D H and Chun I, 1996, *Nanotechnology,* **7**, 216–23.
[3] Greiner A, and Wendorff J H, 2007, *Angew. Chem. Int. Ed.*, **46**, 5670–5703.
[4] MacDiarmid A G, Jones W E, Norris I D, Gao J, Johnson A T, Pinto N J, Hone J, Han B, Ko F K, Okuzaka H, and Llaguno M, 2001; *Synthetic Metals*, **119**, 27–30.
[5] Reneker D H, and Yarin A L, 2008, *Polymer* **49**, 2387-2425.
[6] Chronakis I S, Grapenson S and Jakob A, 2006, *Polymer*, **47,** 1597–1603.
[7] Kang T S, Lee S W, Joo J and Lee J Y, 2005, *Synthetic Metals*, **153**, 61–64.
[8] S. Nair, E. Hsiao and S. H. Kim, J. Mater. Chem., 2008, 18, 5155–5161
[9] Langley P J, Davis F J, and Mitchell G R, 1997, *J. Chem. Soc, Perkin Transactions,* 2229-40.
[10] Sessler J L, Mozaffari A, and Johnson M R, 1998, *Organic Synthesis*, *Coll.Vol.* **9**, p242; 1992, **70**, 68.
[11] Analytical data for 3,4-diethylpyrrole, 3,4-Diethylpyrrole: ^{1}H *NMR* (250 MHz, CDCl3) δ: 1.21 (t, 6 H, CH_3), 2.47 (q, 4 H, CH_2), 6.48 (d, 2 H, pyrrole CH), 7.75 (s, 1 H, pyrrole NH); ^{13}C *NMR* ($CDCl_3$): δ(ppm) = 124.98, 115.02, 32.16, 32.22, 18.95, 18.91, 18.18, 15.06, 14.66; *MS* m/e (relative intensity) 123 (100), 108 (70), 130 (65); *bp* 100°C/25 mm (water aspirator); 69°C/2 mm (vacuum pump) (lit.ref bp, 83°C/10 mm). ν_{max} (DCM film)/cm^{-1} = 3393, 2957, 2916, 2877, 1674, 1456, 1456, 1373, 1067, 943, 782.
[12] Hamlett C A E, Jayasinghe S N and Preece J A, 2008, *Tetrahedron,* **64**, 8476–83
[13] Barrett K E J, 1975, *Dispersion Polymerisation in Organic media* (Wiley, New York).
[14] Kemp M, 1981, *J. Chem. Educ.*, **58**, 655-6.

Acknowledgements

The scanning electron microscopy and x-ray analysis was performed in the Centre for Advanced Microscopy at the University of Reading and we thank the staff for the assistance with the measurements. Sujat Sen thanks the Felix scholarship award scheme for financial support.

Structure development in electrospun fibres of gelatin

Delyth E Elliott, Fred J Davis, Geoffrey R Mitchell[1] and Robert H Olley
Polymer Science Centre, University of Reading, Whiteknights,
Reading, RG6 6AF

E-mail: g.r.mitchell@rdg.ac.uk

Abstract. Gelatin fibres have been successfully electrospun from water by heating a gelatin solution above the sol-gel transition temperature, and allowing cooling in a controlled environment as the fibres are produced. The development of structure with in these fibres is monitored using wide angle x-ray scattering, in this way the presence of the triple helix structure, which provides the physical cross-linkages in the gel could be probed. There is clear evidence that these structures are obtained in gelatin electrospun from aqueous solutions. In contrast fibres electrospun from a solution of gelatin in glacial acetic acid, showed no evidence of the triple helix structure.

1. Introduction

Electrospinning is a technique for producing nanoscale fibres from a solution with the aid of an electric field [1]. This study looks into the gelling process which can be found in biopolymers such as gelatin in aqueous solutions; and its effect on the electrospinning process. The production of nanofibres is an essential step in the optimisation of a range of biomedical and pharmaceutical applications. These include tissue engineering, drug release, and wound dressings [2]. Gelatin, with its inherent biocompatibility, is an attractive material choice for such applications. Unfortunately, until now gelatin has only been electrospun when dissolved in selected non-aqueous solvents; these include 2,2,2-trifluoroethanol, glacial acetic acid and acetic acid / dimethyl sulfoxide [3], [4]. The advantage of making a nanofibre from water is that the product would be non-toxic, an attractive option for biomedical applications.

Gelatin is a complex biopolymer system containing principally glycine, proline and 4-hydroxyproline [5]. These amino acids are bonded together forming a polymer chain, the state of which is dependent on temperature. In aqueous solution it exhibits a gel at temperatures below ~ 30 °C. The gelation is driven by hydrogen bonding, and in particular, the formation of short sequences of triple helix structures similar to that observed in collagen. Above this temperature the triple helix structure partially unravels reducing the number of hydrogen bonds, and resulting in a change of state from gel to solution [5]. The generation of nanoscale gelatin fibres using electrospinning requires the destabilisation of the triple helix structures. This can be achieved through the use of an acidic solvent or through the hitherto unexplored approach of electrospinning a solution of gelatin and water at an elevated temperature, which is described here for the first time.

[1] Author to whom enquiries should be addressed

2. Experimental Procedure

2.1 Materials

In this work the gelatin used was bovine gelatin sourced from Gelita with a Bloom value of 210. All chemicals and samples were obtained commercially and used without further purification, except for water which was distilled prior to use. Solutions were made by shaking the required quantity of solid with the hot solvent until all the solid was observed to have dissolved.

2.2 Electrospinning

Electrospinning was performed as shown in Figure 1. A glass syringe mounted in a syringe pump was fitted with a 22 gauge needle of length 2.5 cm (and with an internal diameter of 0.71 mm). A flat aluminium electrode was placed normal to the needle at distances varying from 10 to 40 cm. A Glassman High Voltage power supply was used which allowed defined voltages over the range 7.5-20 kV. In these experiments (unless otherwise stated) a potential of 13kV was used with a needle to collector distance of 14 cm. The temperature of the solution was maintained at the desired temperature (Ts) using a heating stage, while an air conditioning unit provided some control of the ambient temperature (Ta) within an environmental chamber.

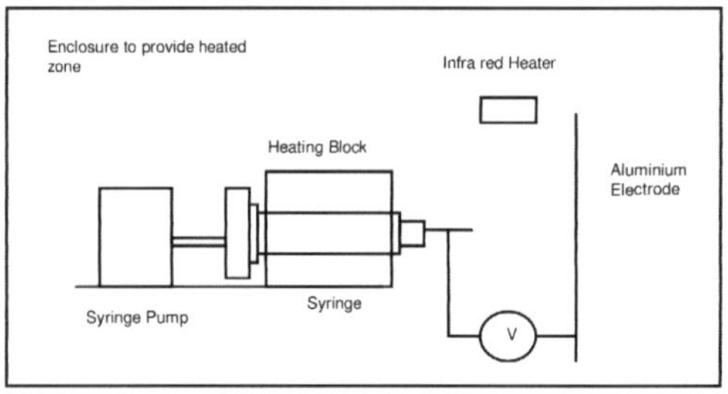

Figure 1. The experimental set-up.

2.3 Wide angle x-ray scattering (WAXS)

WAXS data on fibre samples at 5 °C over an extended |Q| range were obtained using a symmetrical transmission diffractometer equipped with a graphite monochromator and pinhole collimation and a Cu K source (with a wavelength of 1.54 Å). The intensity values from samples were obtained as a function of |Q| over the range 0.2 to 6 $Å^{-1}$ in steps of 0.02 $Å^{-1}$ where $|Q| = 4\pi\sin\theta/\lambda$, 2θ is the scattering angle and λ is the incident x-ray wavelength. The fibres were held on a temperature controlled plate at 5 °C to ensure they did not melt (and hence retained their molecular structures).

3. Results

3.1 Electrospinning fibres from a solution of gelatin and acetic acid.

Gelatin was dissolved in acetic acid to form a 30% w/v concentration solution. This was then electrospun to produce smooth fibres, without beading, as shown in the SEM image in figure 2a. The fibres produced had an average fibre diameter of approximately 1000 nm, which is of a similar order of magnitude to the fibres produced by Choktaweesap *et al* [4]. In order to establish the presence of any triple helices, similar fibres were produced for a WAXS experiment, the results of which are shown in figure 2b. These show a disordered polymer with no peaks at Q = 2.2 $Å^{-1}$, which are characteristic of a triple helix when gelatin is dissolved in water and in the gel state as shown by Bigi *et al* [6].

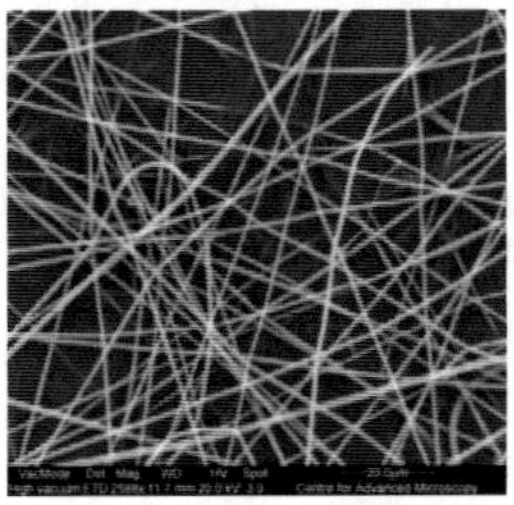

Figure 2a. Gelatin electrospun fibres prepared from acetic acid solution with Ts = 25 °C

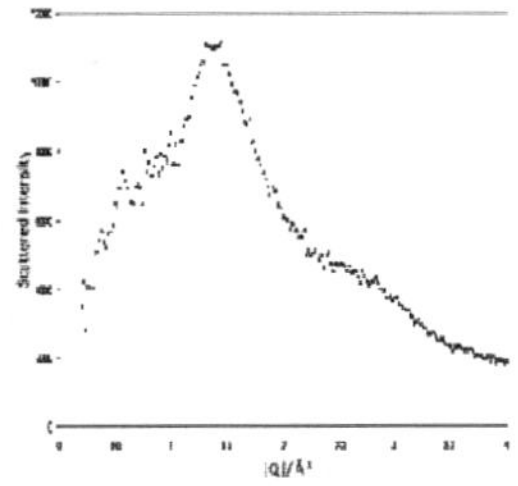

Figure 2b. Wide-angle x-ray scattering curve for fibres prepared from acetic acid solution with Ts = 25 °C

3.3 Electrospinning fibres from a solution of gelatin and water.

We have developed a methodology which allows gelatin fibres to be spun directly from aqueous solutions by holding the solution at an elevated temperature. The air temperature is adjusted to optimise the process, controlling the rate of gelation to facilitate electrospinning. In this way fibres can be directly electrospun from an aqueous gelatin solution as shown in figure 3a. The fibres were free from beading and essentially cylindrical; with variable diameters depending on the conditions chosen. A typical distribution of fibre diameters is shown in figure 3b.

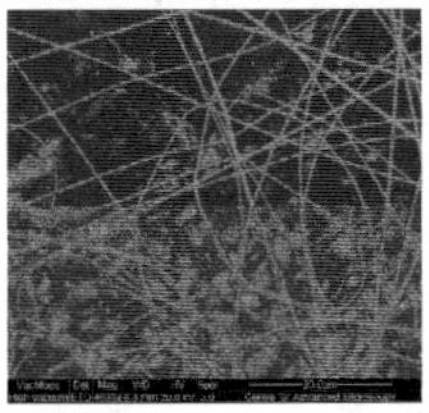

Figure 3a. Gelatin electrospun fibres prepared from aqueous solution with Ts = 49 °C and Ta = 33 °C

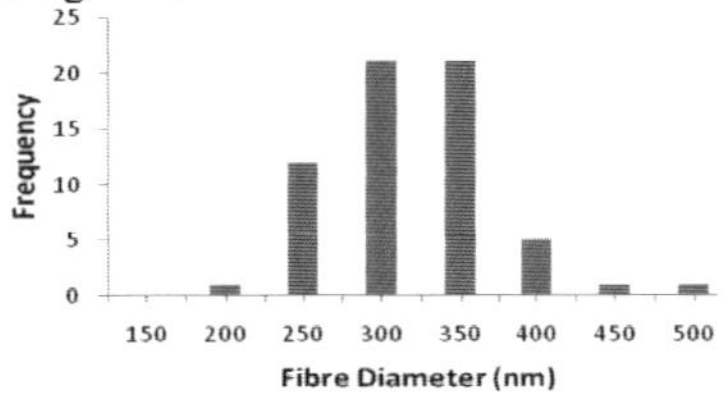

Figure 3b. A histogram showing the variation of fibre diameter for fibres spun at 37 °C ± 3 °C. The average fibre diameter is 295 nm.

It was found that electrospinning aqueous gelatin fibres is not possible for Ts < 36 °C; at these temperatures presumably the solutions become too viscous. A general trend is observed for the fibres prepared at different temperatures, in that an increasing diameter is coupled with increasing values of Ts as shown in Table 1. This can be correlated with the change in viscosity, and the greater delay in any gelation. As a comparison, gelatin fibres electrospun from a 30% w/v acetic acid solution using similar experimental conditions, other than the temperature, Ts = 25 °C, generated fibres with a diameter of ~ 1000 nm.

Table 1. Fibre diameter as a function of Ts at which the fibres were electrospun.

Ts (°C)	*Average Fibre Diameter (nm)*
25	no fibres
37	295
43	269
49	306
52	376

3.2 Structure and Gelation

Triple helices found within the collagen molecular structure have been identified by x-ray diffraction by their characteristic diffraction rings. The rings correspond to two distances, 11 Å and 2.9 Å. The first refers to the diameter of the triple helix and the second relates to the distance between the amino acid residues along the helix [6]; thus we analysed the fibres with a view to monitoring the x-ray scattering at values for |Q| of 0.571Å^{-1} and 2.17 Å^{-1}. As an initial probe of the way in which such helices form, the WAXS curves (Figure 3) were obtained at room temperature from gelatin and water samples cast at the temperatures shown and the water removed by reduced pressure. The sharper peaks at ~ 0.5 Å^{-1} and 2.2 Å^{-1} are a direct signature of the formation of the triple helix structures; materials prepared at temperatures above the sol-gel transition temperature do not exhibit this feature.

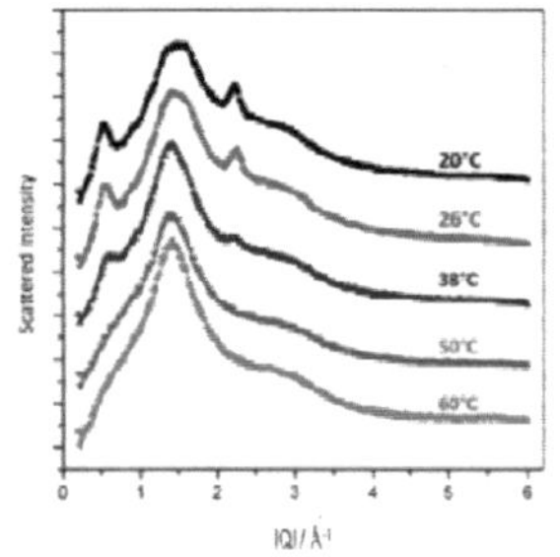

Figure 4. Wide-angle x-ray scattering data recorded for gelatin samples cast from aqueous solutions at the temperatures indicated and dried at that temperature.

3.4 Structure in aqueous gelatin fibres

Unlike those fibres spun from acetic acid, (Figure 2b) the wide-angle scattering of the aqueous gelatin fibres shows that there is significant triple helix content. For example, Figure 5 shows the WAXS of fibres spun with Ts = 45 °C and 53 °C although the triple helix content is significant, it is rather less than that exhibited by a room temperature gel. Interestingly the features arising from the triple helix appeared stronger both at a higher spinning temperature, and when the fibres were left to dry.

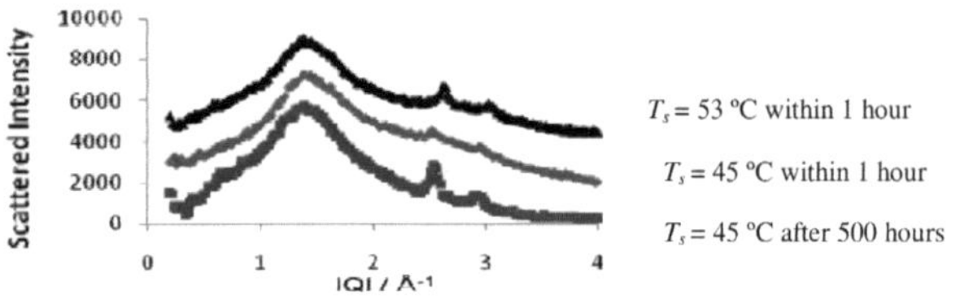

Figure 5. Wide-angle x-ray scattering data for gelatin fibres prepared by electrospinning from aqueous solution with Ts = 45 °C, 53 °C and Ta = 36 °C.

4. Conclusions

Nanoscale fibres of gelatin can be directly prepared from aqueous solution using thermal electrospinning techniques for Ts>36°C. Fibres prepared from aqueous solution exhibit a size distribution which is dependent on the temperature of the electrospinning solution. Wide-angle x-ray scattering shows that these fibres prepared by this thermal electrospinning process show some triple helix content, but less than that observed in the equivalent room temperature gelatin gels. Gelatin fibres prepared from acetic acid solution at 25°C show no triple helix content. The successful production of non-toxic gelatin nanofibres provides an additional route to biomedical applications.

5. References

[1] Ramakrishna S, Fujihara K, Teo W E, Lim T C and Ma Z., 2005, *An Introduction to Electrospinning and Nanofibers* (World Scientific Publishing Co Pte Ltd)
[2] Agarwal S, Wendorff J H and Greiner A., 2008, *Polymer* **49** 5603 – 5621.
[3] Huang Z M, Zhang Y Z, Ramakrishna S and Lim C T. 2004 *Polymer* **45** 5361 – 5368
[4] Choktaweesap N, Arayanarakul K, Aht-ong D, Meechaisue C, and Supaphol P. 2007 *Polymer Journal* **39** 622 – 631
[5] Schrieber R and Gareis H. 2007 *Gelatine Handbook Theory and Industrial Practice* (Wiley-VCH)
[6] Bigi A, Panzavolta S and Rubini K. 2004 *Biomaterials* **25** 5675–5680

Acknowledgements

This work was funded by the EPSRC, the Deutsches Elektronen-Synchrotron (DESY) and the European Community (RII3-CT-2004-506008 (IA-SFS)). We thank Dr Sérgio Funari for his assistance with the x-ray scattering measurements on the A2 beam-line at DESY. We thank Gelita UK Ltd for the gift of the gelatin.

Isolation and measurement of the features of arrays of cell aggregates formed by dielectrophoresis using the user-specified Multi Regions Masking (MRM) technique

Rama Yusvana*, Denis Headon†, and Gerard H. Markx*

*School of Engineering and Physical Science, Department of Chemical Engineering, Heriot-Watt University, Riccarton Campus, Edinburgh - EH14 4AS, UK

†The Roslin Institute and Royal (Dick) School of Veterinary Studies, University of Edinburgh, Roslin, Midlothian, Edinburgh - EH25 9PS, UK

g.h.markx@hw.ac.uk

Abstract. The use of dielectrophoresis for the construction of artificial skin tissue with skin cells in follicle-like 3D cell aggregates in well-defined patterns is demonstrated. To analyse the patterns produced and to study their development after their formation a Virtual Instrument (VI) system was developed using the LabVIEW IMAQ Vision Development Module. A series of programming functions (algorithms) was used to isolate the features on the image (in our case; the patterned aggregates) and separate them from all other unwanted regions on the image. The image was subsequently converted into a binary version, covering only the desired microarray regions which could then be analysed by computer for automatic object measurements. The analysis utilized the simple and easy-to-use User-Specified Multi-Regions Masking (MRM) technique, which allows one to concentrate the analysis on the desired regions specified in the mask. This simplified the algorithms for the analysis of images of cell arrays having similar geometrical properties. By having a collection of scripts containing masks of different patterns, it was possible to quickly and efficiently develop sets of custom virtual instruments for the offline or online analysis of images of cell arrays in the database.

1. Introduction

During embryonic development of skin tissue in chicken feather bud primordia are formed in regular hexagonal patterns. Patterning starts at a line above the neural tube, which subsequently breaks up into individual placodes. Formation of placodes then spreads from this line in a cascade over the whole skin, forming a hexagonal array of feather buds. The exact details of the mechanisms by which this pattern is formed are still unknown, but are thought to involve the condensation of cells into aggregates (placodes) whose position is determined by a reaction-diffusion mechanism of activator and inhibitor molecules distributing themselves at differential rate across the skin [1].

The creation of artificial skin with cell aggregates of known size and shape spaced in well defined and regular patterns could help with elucidating the mechanisms by which the regular arrays of feather bud primordia are formed. In this study the use of dielectrophoresis (DEP), the lateral movement of particles in non-uniform electric fields [2] was explored for the formation of cell aggregates of controlled size and shape in well-defined patterns.

After the formation of the aggregates it is often necessary to analyze the patterns produced and to study their development after their formation. To aid in this we have developed LabVIEW-based Virtual Instrument (VI) system to effectively and simultaneously analyze images of microarrays of cell aggregates using IMAQ Vision Development Module, which we will describe here.

2. Methods

Electrode designs were made using Protel Design Explorer 99. Electric field calculations were made using COMSOL Multiphysics. Microelectrodes were made from ITO-covered glass slides using standard photolithography. A simple microfluidic chamber was created above the microelectrode array using glass slides and a silicone spacer (see Figure 1). Chicken embryonic skin cells were isolated by enzymatic tissue digestion from the dorsal part of an 8-day old white leghorn chicken embryo. The dissociated skin cells were suspended in Dulbecco medium (DMEM) prior to DEP experiments. When needed, the cells were centrifuged and washed several times before transferring to low conductivity iso-osmotic sorbitol solution (300 mM, $\sigma = 6.1 \times 10^{-4}$ S m^{-1}). The cells were introduced into the DEP chamber (approximate height = 500 μm) at a constant flow rate of around 40 μL minute^{-1} using a peristaltic pump.

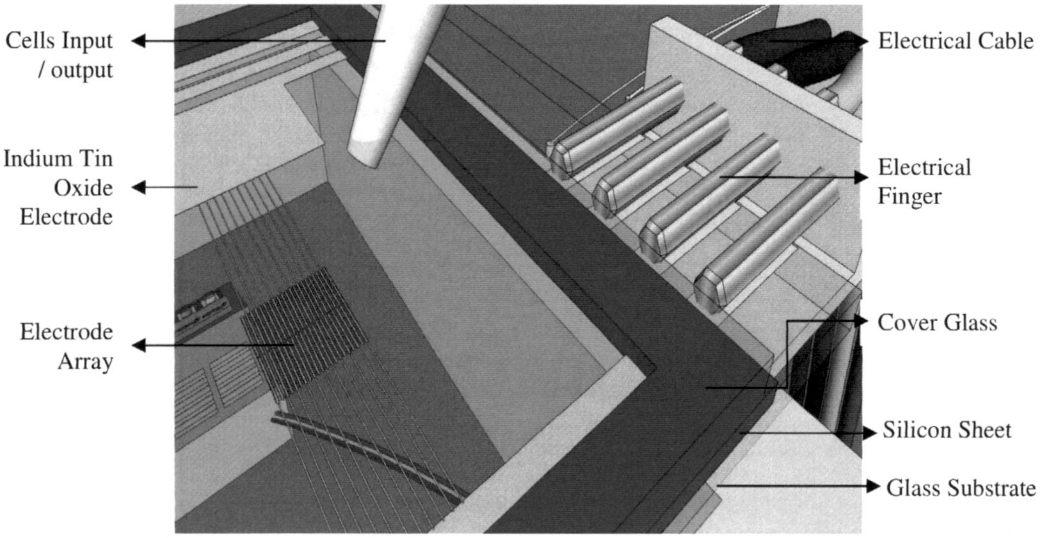

Figure 1. Dielectrophoresis chamber for the formation of well defined biodielectric patterns.

Construction of the biodielectric patterns involved the application of a high frequency (1 MHz, 10 $V_{pk\text{-}pk}$) AC signal using TG120 Thurlby-Thandar frequency generator, to the microelectrodes to create non-uniform electric fields and use dielectrophoresis to force the cells together. This was followed by the immobilization of the cells in a 50% Puramatrix gel in order to create a skin-like material and enable the long-term incubation of the immobilized cells. Images were taken of the aggregates using a widefield stereo microscope from Celsi and a NIKON Coolpix E4500 digital camera. Image analysis was performed with LabVIEW IMAQ Vision.

3. Formation of Various and Defined Biodielectric Patterns

As shown previously [3] the shape of cell aggregates between a set of castellated electrodes can be predicted using standard finite element analysis package such as COMSOL Multiphysics. The shape of the cell aggregates could be expected to follow the pattern of the high electric field strength regions between castellations.

Electric field patterns formed between electrodes of different shapes are shown in Figure 2. 2D images of the actual aggregates of skin cells formed and their associated pseudo-3D representations based on the distribution of the pixel intensities are shown in Figure 3(I) and 3(II) respectively. Although the shape of the cell aggregation followed the predicted pattern based on electric field calculations well, it was affected by factors such as the liquid flow rate and the concentration of cells entering the chamber, as well as the electrode geometry, frequency and potential difference between the castellations.

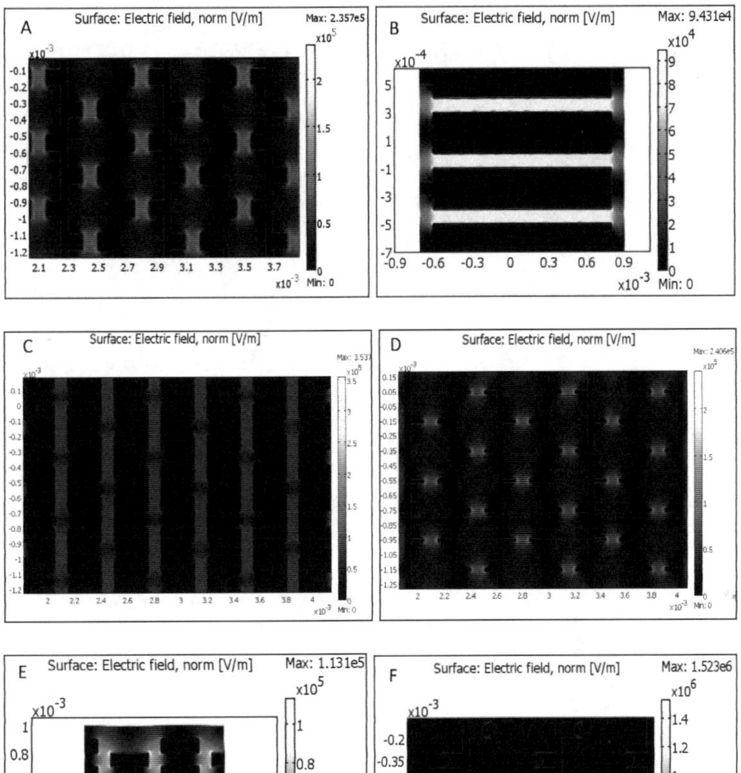

Figure 2. Modelling of electric field calculation between sets of castellated electrodes with various design patterns (in meter unit). A. Small Hexagon. B. Line Pattern. C. Large Hexagon. D. Elongated Hexagon. E. 'W' Pattern. F. Round Edge with Hexagonal Pattern.

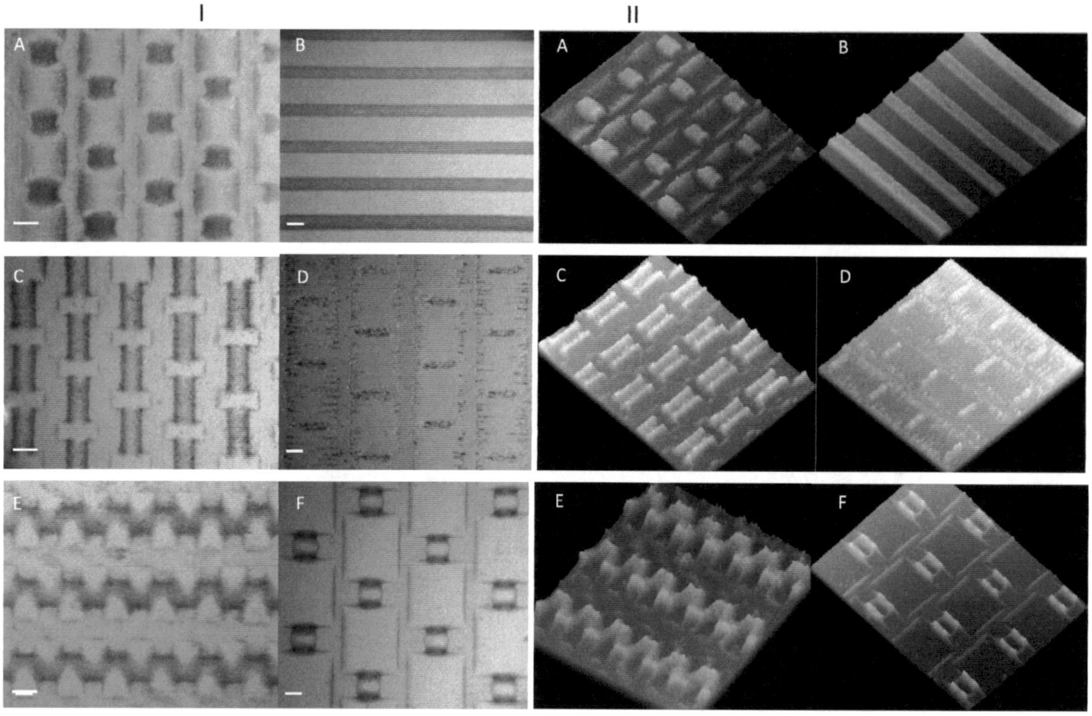

Figure 3. I. Images of embryonic skin cell aggregates formed with DEP in various different patterns. II. 3D representation of cell aggregates between electrode castellation based on pixel intensity. A. Small Hexagon. B. Line Pattern. C. Large Hexagon. D. Elongated Hexagon. E. 'W'Pattern. F. Round Edge with Hexagonal Pattern. Scale bar is 100 μm.

4. Analyzing the Images Using User-Specified Multi Regions Masking Technique

Analysis of the images of cell aggregates could provide information about - amongst others - the quality of the aggregates, the aggregation process itself, and the development of the aggregates after their formation. To achieve this, it is necessary to take a series of graphical programming steps to convert the original image to binary version showing only the regions of interest (in our case; cell aggregates in the regions specified by the castellations), based on a certain threshold value (illustrated in Figure 4). Once the correct (representative) binary version of the particles is achieved, subsequent analysis is relatively straight forward.

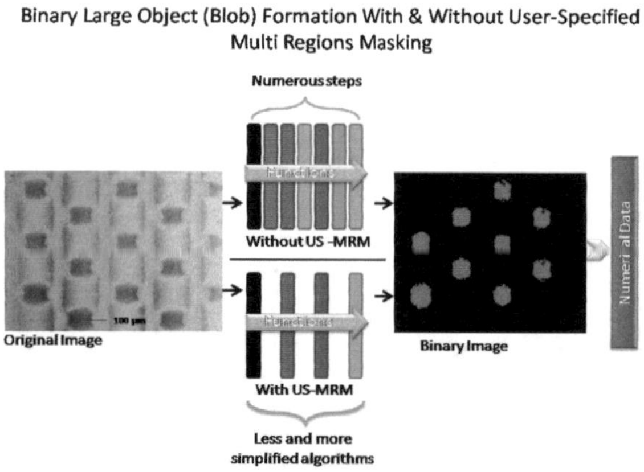

Figure 4. Diagram illustrating differences in graphical programming steps from original image to the formation of binary version of the image.

A series of graphical programming functions (algorithms) was tested and optimized in order to develop Virtual Instruments (VIs) which are capable of analyzing images of arrays of aggregates in a database by simply changing the set parameter values according to the unique conditions of the individual images (e.g. light intensity, edge sharpness). Due to the complex processes that occur during DEP pattern formation and the large number of potential sources of interference (such as unbound cells, contaminating cell debris or other particles), the isolation and measurements of distinct features (immobilized cell aggregates) in the image was complicated. This can be clearly seen in Figure 3, where cells can be seen to aggregate in areas outside the main areas of interest as local high electric field regions also occur outside the main areas [4]. The problems were minimized by using the user-specified multi-regions masking technique prior to binarizing the image for subsequent automatic particle measurements. A similar masking approach was previously used by Gascoyne et al. [5] for measuring aggregate formation with dielectrophoresis. However, their interest was mainly in the speed of the particles forming the aggregates, whilst our interest is mainly in the shape, size and position of the aggregates.

It was found that the best method of specifically isolating only the cell aggregates for further analysis involved applying a set of user-specified multi-region masks prior to thresholding the images based on other parameter values. Thus, image analysis using LabVIEW IMAQ Vision mainly involved the following major group of functions [6]:

- **Image Acquisition** (Algorithm used: *Image open*). This step could be performed *online / real-time* using various types of image acquisition [7] hardware or *offline* (from large collection of images previously saved in a database). In this work only the *offline* method of image analysis was used.
- **Pre-Processing**. This step is where multi-regions masking technique was applied (Algorithm used: *Image Masking*). A series of maskings was applied on the displayed image using the various masking tools available in the package (*circular / oval*, *square*, *polygon*, *annulus*, *rotated rectangle,* or *freehand region* tool). This was achieved by holding Ctrl button (in PC) during multi-regions masking. The mask covered or filtered out the majority of the unwanted regions in the image, leaving only the regions of interest for further analysis (see Figure 5-I). Other algorithms that could be used in this pre-processing step are *Calibration* (to give real-life metric measurements instead of pixel measurements), and *Plane Extraction* (to give grayscale version of the original full color image).
- **Thresholding**. This step isolated the pixels of interest in the previously unmasked areas (to be a set value = 1) and set the remaining pixels as background pixels (set value = 0) thereby creating a binary image. A locally adaptive thresholding method (*Niblack*) was used which categorized a pixel as part of particle or the background based on the intensity statistics of its neighboring pixels. This method was particularly good when the original images exhibited non-uniform lighting effect during acquisition.
- **Filtering**. Because thresholding is a subjective process, the resulting binary image may still contain unwanted information, such as small residual noise particles. Use of the Multi-regions masking technique cut out the majority of these noises. The remaining filtering algorithms used were therefore *Remove small objects* and *Remove border objects* (i.e. the mask itself). Other optional algorithms that were sometimes needed fell under binary morphology modification functions. These included *Fill holes, Close objects,* and *Dilate / Erode objects*. The final binary images formed the closest possible representative shapes of the actual biodielectric patterns (Figure 5-II).
- **Measurement**. IMAQ Vision *Particle Analysis* could be used to calculate all (up to 49) requested measurements in pixel or metric values. Measurements used here included object perimeter, area, hydraulic radius, elongation factor and Heywood Circularity factor (Table 1 and Table 2).

The final step was the generation of custom VIs, based on selected algorithms above. This was used to analyze biodielectric images requiring similar mask templates and the same geometrical pattern. The same script could be used for batch processing other images in a library [8]. However, adjustment of the threshold value was often required for a more representative binary version of the original image.

The calculations performed were relatively fast. All steps taken to analyze all objects simultaneously in each image (including isolation and measurement processes) could be implemented efficiently with a total time less than 0.5 seconds, on a 2 GHz Pentium 4 processor.

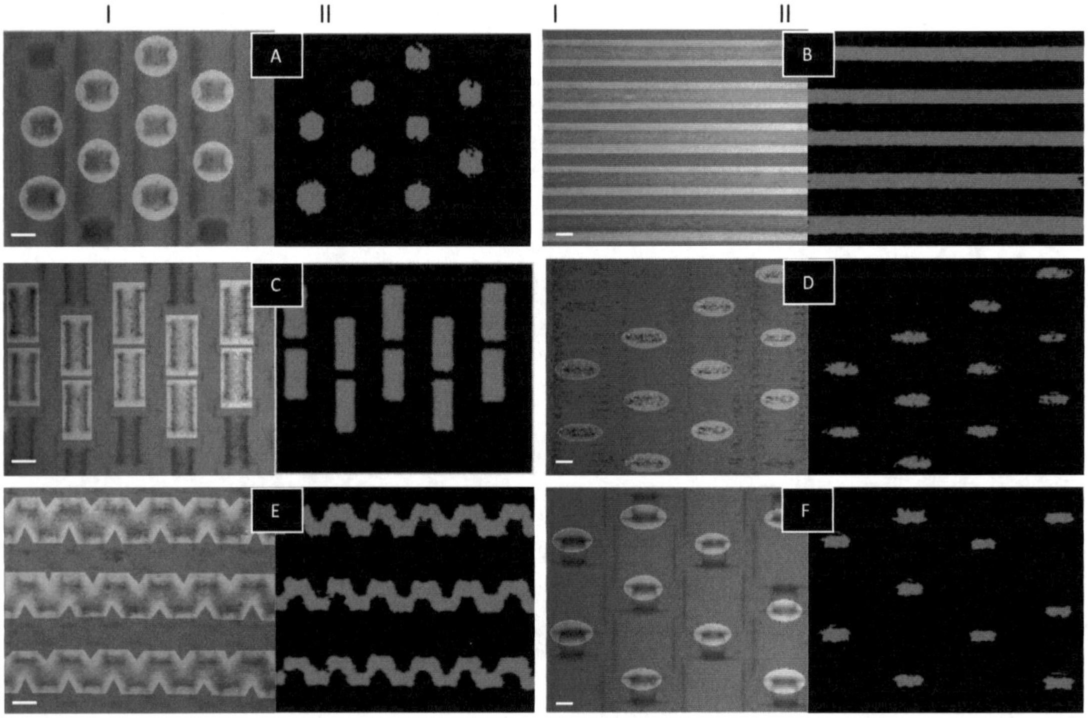

Figure 5. Conversion of user-specified masked images (I) to binary large objects (II), a group of contiguous pixels that have the same intensity value (i.e. 1). A. Small Hexagon. B. Line Pattern. C. Large Hexagon. D. Elongated Hexagon. E. 'W'Pattern. F. Round Edge with Hexagonal Pattern. Scale bar is for 100 µm.

5. Results of Image Analysis and Discussion

The results of the analysis of the images shown in Figures 3 and 5 using the user-specified multi regions masking technique described here are given in Tables 1 and 2.

Table 1 confirms that the masking technique can be used to generate data from images of cell aggregates formed at a variety of electrode shapes. Table 2 shows a comparison of the distances and angles between cell aggregates and compares them with the expected values based on the specifications of the actual photo mask design. The analysis shows a very good agreement between the two sets of data.

Table 1. Summary of 2D geometric measurements / analysis of aggregates on various electrode designs. All electrodes had a characteristic size of 100 μm. Values given are the mean and standard error of mean

Microelectrode Designs	Mask Type	Perimeter (μm)	Elongation Factor	Circularity Factor	Area (μm^2)	Hydraulic Radius
Small Hexagon	Circular	290.1 ± 10.12	1.97 ± 0.04	1.41 ± 0.03	3410.2 ± 197.2	11.71 ± 0.42
Line Pattern	Square	4772.2 ± 101.2	10.53 ± 0.51	2.72 ± 0.08	247428.7 ± 14917.6	51.85 ± 2.80
Large Hexagon	Square	1119.1 ± 10.3	2.43 ± 0.03	1.38 ± 0.02	52590.7 ± 970.2	47.03 ± 1.06
Elongated	Oval	525.2 ± 10.3	3.21 ± 0.10	1.56 ±0.03	9067.2 ± 363.1	17.26 ± 0.62
'W' Pattern	Polygon	9320.1 ± 587.9	N/A	N/A	223079.2±7683.2	N/A
Round Edge	Oval	495.9 ± 21.8	3.38 ± 0.21	1.73 ±0.09	8993.2 ± 197.4	18.62 ± 0.97

N/A = Not Available

Table 2. Summary of measured distances and angles between cell aggregates formed at different electrode designs and the expected value based on the electrode design. All electrodes had a characteristic size of 100 μm. Values given are the mean and standard error of mean.

Microelectrode Designs	Vertical Distances (μm)	Exp. Value (μm)	Horizontal Distances (μm)	Exp. Value (μm)	3-Points Angle 1 (°)	Exp. Value (°)	3-Points Angle 2 (°)	Exp. Value (°)
Small Hexagon	393.2 ± 1.2	400	442.7 ± 1.3*	450*	115.2 ± 1.1	116.4	124.8 ± 2.3	127.3
Line Pattern	402.0 ± 5.0	400	N/A	N/A	180.1 ± 1.1	180	N/A	N/A
Large Hexagon	397.8 ± 3.6	400	803.5 ± 5.8	800	119.1 ± 2.7	120	122.5 ± 5.1	120
Elongated	401.7 ± 2.8	400	799.5 ± 3.9	800	121.9 ± 4.1	120	119.7 ± 4.8	120
'W' Pattern	403.6 ± 6.6	400	N/A	N/A	181.4 ± 3.7	180	N/A	N/A
Round Edge	398.4 ± 2.9	400	807.1 ± 5.9	800	120.8 ± 1.9	120	121.3 ± 2.9	120

*Diagonal data shown. Exp. = Expected Value. N/A = Not Available

It should be noted that the multi regions masking technique is invariant to the scale, shape and rotation of the features as the users can specify the region just outside the features of interest. The technique is therefore highly useful in the sense that a single mask design can correctly isolate other features (from other images in the database) within the unmasked area with only a few threshold level adjustments, depending on the variation in illumination level and camera viewpoint during acquisition.

Successful isolation of features is often a key and fundamental aspect of many problems in computer vision and image analysis. The multi regions masking technique described here is particularly useful due to their reusability with only a few threshold level adjustments. Furthermore, statistical analysis can be performed on the data to determine whether the changes are significant or not.

There are many directions for further research in which algorithm development includes addition of recognition feature of biodielectric patterns, based on either shape or in combination with their distinctive color. This includes algorithm development which aimed to individually learn the repeating structure (individual cell aggregate) which will then be used to recognize and subsequently analyze other object within the same group or categories. Some of these developments may combine or utilize the shape recognition technique with edge-based features [9].

6. Conclusions

It has been shown that dielectrophoresis can be used to create arrays of cell aggregates in well-defined patterns. By controlling the design of the microelectrode geometry it is possible to manipulate the 2D and 3D geometry of the resulting cell aggregates. This control includes their spacing (distance and angle between the aggregates), their size and shape as well as cell density distribution. Control of the timing of the introduction of different cells entering the chamber (e.g. dermal cells followed by epidermal cells on top) allows one to form distinct multi-layer micro niches within a controlled microenvironment [10], making it possible to use the system as a tool for the study of the interactions between cells and their microenvironment [11]

It has also been shown that the Multi Regions Masking technique greatly simplify sets of graphical programming (functions) used in image analysis. It allows users to focus directly on the region of interest (ROI) and perform various numerical measurements only on the desired regions (objects). LabVIEW IMAQ Vision Assistant facilitates the development of custom Virtual Instruments for subsequent batch image processing with only a few (minor) threshold level adjustments. The custom VI development is particularly useful if combined with a permanent imaging setup (giving minimum variation in illumination level between acquisition batches). The results of image analysis performed here and the geometry of electrode design used have shown excellent agreement in terms of expected distance, angle and orientation of the desired features.

7. References

[1]. Mou C, Jackson B, Schneider P, Overbeek PA and Headon DJ. Generation of the primary hair follicle pattern. Proc. Natl. Acad. Sci. USA. 2006;103:9075-9080.
[2]. Markx GH. The use of electric fields in tissue engineering. Organogenesis 2008;4:11-17.
[3]. Sebastian A, Venkatesh AG and Markx GH. Tissue engineering with electric fields: Investigation of the shape of mammalian cell aggregates formed at interdigitated oppositely castellated electrodes. Electrophoresis 2007; 28:3821-3828.
[4] Wang XB, Huang Y, Burt JPH, Markx GH, Pethig R. Selective dielectrophoretic confinement of bioparticles in potential-energy wells. J. Phys. D: Appl. Phys. 1993;26:1278-1285.
[5] Gascoyne PRC, Huang Y, Pethig R, Vykoukal J and Becker FF. Dielectrophoretic separation of mammalian cells studied by computerized image analysis. Meas. Sci. technol. 1992;3:439-445.
[6]. National-Instrument. NI Vision Concepts Manual. National Instrument Manual. 2007.
[7]. National-Instrument. NI Image Acquisition Fundamentals. National Instrument Manual.
[8]. National-Instrument. NI Vision Assistant Tutorial. National Instrument Manual 2007.
[9]. Mikolajczyk K, Zisserman A and Schmid C. Shape recognition with edge-based features. Proceeding of the British Machine Vision Conference, Norwich U.K. 2003.
[10] Sebastian A, Buckle AM and Markx GH. Tissue engineering with electric fields: immobilisation of mammalian cells in multilayer aggregates using dielectrophoresis. Biotechnol. Bioeng. 2007;98:694-700.
[11] Markx GH, Carney L, Littlefair M, Sebastian A and Buckle AM. Recreating the hematon: microfabrication of artificial haematopoietic stem cell microniches in vitro using dielectrophoresis. Biomedical Microdevices 2009;11:143-150.

AUTHOR INDEX

Institute of Physics
Dirac House, Temple Back
Bristol BS1 6BE UK